Planning, Law and Economics

Planning, Law and Economics sets out a new framework for applying a legal approach to spatial planning, showing how to improve the practice and help achieve its aims. The book covers planning laws, citizens' rights and property rights, asking 'What rules do we want to make and, where necessary, enforce? And how do we want to apply them in planning practice?' This book sets out, in general and illustrated with concrete examples, how the three types of law mentioned above are unavoidably involved in all types of spatial planning. The book also makes clear that these laws can be combined in different ways, each way a particular approach to the practice of spatial planning (regulative planning, structuring markets, pro-active planning, collaborative planning, etc.).

Throughout, the book shows what legal approaches can be taken to spatial planning, and uses a four-part framework to evaluate the effects of choosing such an approach. The spatial planning should be effective, legitimate, morally just and economically sound. In particular the book details why the economic effects for society are important and how spatial planning affects how the economic resources of land and buildings are used. The book will be invaluable to students and planners to understand the relationship between their actions and the basic principles of the rule of law in a democratic, liberal society.

Barrie Needham is emeritus professor of spatial planning, Radboud University, Nijmegen, the Netherlands.

Edwin Buitelaar is a professor of land and real estate development at Utrecht University, the Netherlands, senior researcher on urban development at the Netherlands Environmental Assessment Agency (PBL), and research fellow at the Amsterdam School of Real Estate (ASRE).

Thomas Hartmann is associate professor at the Land Use Planning Group of Wageningen University & Research, the Netherlands, and affiliated to the Faculty of Social and Economic Studies of the JEP University (UJEP) in Ústi nad Labem, Faculty of Social and Economic Studies, Czech Republic.

T0138792

THE RTPI Library Series

Editors: Robert Upton, *Infrastructure Planning Commission in England*
Jill Grant, *Dalhousie University, Canada*
Stephen Ward, *Oxford Brookes University, United Kingdom*

Published by Routledge in conjunction with The Royal Town Planning Institute, this series of leading-edge texts looks at all aspects of spatial planning theory and practice from a comparative and international perspective.

The Craft of Collaborative Planning: People Working Together to Shape Creative and Sustainable Places
Jeff Bishop

Future Directions for the European Shrinking City
Edited by Hans Schlappa and William J.V. Neill

Insurgencies and Revolutions
Edited by Haripriya Rangan, Kam Mee NG, Libby Porter and Jacquelyn Chase

Planning for Small Town Change
Neil Powe and Trevor Hart

Regent Park Redux: Reinventing Public Housing in Canada
Laura Johnson and Robert Johnson

Planning in Indigenous Australia: From Imperial Foundations to Postcolonial Futures
Sue Jackson, Libby Porter and Louise C. Johnson

From Student to Urban Planner: Young Practitioners' Reflections on Contemporary Ethical Challenges
Edited by Tuna Taşan-Kok and Mark Oranje

Public Norms and Aspirations: The Turn to Institutions in Action
Willem Salet

Planning, Law and Economics: The Rules We Make for Using Land, 2nd edition
Barrie Needham, Edwin Buitelaar and Thomas Hartmann

Planning, Law and Economics

The Rules We Make for Using Land

Second Edition

Barrie Needham, Edwin Buitelaar
and Thomas Hartmann

Routledge
Taylor & Francis Group

NEW YORK AND LONDON

Second edition published 2019
by Routledge
52 Vanderbilt Avenue, New York, NY 10017

and by Routledge
2 Park Square, Milton Park, Abingdon, Oxon, OX14 4RN

Routledge is an imprint of the Taylor & Francis Group, an informa business

© 2019 Taylor & Francis

The right of Barrie Needham, Edwin Buitelaar and Thomas Hartmann
to be identified as authors of this work has been asserted by them in
accordance with sections 77 and 78 of the Copyright, Designs and
Patents Act 1988.

All rights reserved. No part of this book may be reprinted or reproduced
or utilised in any form or by any electronic, mechanical, or other
means, now known or hereafter invented, including photocopying and
recording, or in any information storage or retrieval system, without
permission in writing from the publishers.

Trademark notice: Product or corporate names may be trademarks or
registered trademarks, and are used only for identification and
explanation without intent to infringe.

First edition published by Routledge 2006

Library of Congress Cataloging-in-Publication Data
Names: Needham, Barrie, author. | Buitelaar, Edwin, author. |
    Hartmann, Thomas, 1979- , author.
Title: Planning law and economics : the rules we make for using land /
    Barrie Needham, Edwin Buitelaar and Thomas Hartmaan.
Description: Second edition. | Milton Park, Abingdon, Oxon ;
    New York, NY : Routledge, 2018. | Series: RTPI library series |
    Includes bibliographical references and index.
Identifiers: LCCN 2018009037| ISBN 9781138085558 (hardback) |
    ISBN 9781138085572 (pbk.) | ISBN 9781315111278 (ebook)
Subjects: LCSH: Land use—Law and legislation. | Regional
    planning—Law and legislation. | Right of property.
Classification: LCC K3534 .N44 2018 | DDC 346.04/5—dc23
LC record available at https://lccn.loc.gov/2018009037

ISBN: 978-1-138-08555-8 (hbk)
ISBN: 978-1-138-08557-2 (pbk)
ISBN: 978-1-315-11127-8 (ebk)

Typeset in Goudy
by Swales & Willis Ltd, Exeter, Devon, UK

Contents

List of Tables   vii
Preface to the Second Edition   ix
Preface to the First Edition (2006)   xi

1   Why Planning, Law and Economics Matter   1

2   Property Rights in Land and Buildings   25

3   Planning Law   48

4   Citizens' Rights in Spatial Planning   69

5   Law and Policy Effectiveness and Efficiency in
    Spatial Planning   85

6   Law and Economic Welfare in Spatial Planning   97

7   Law and Justice in Spatial Planning   125

8   Law and Legitimacy in Spatial Planning   138

9   Using the Law in Practice   148

    Index   165

Tables

2.1	Types of property regime	42
2.2	Types of rules that influence the exercise of property rights	45
3.1	Two sorts of land-use rules	52
6.1	Private goods, public goods, club goods	103

Preface to the Second Edition

The first edition of this book (2006) investigated two ways of spatial planning – that is, two ways in which a society can try to influence the way in which land is used. One way was by using planning law to regulate how people use their property rights in land and buildings, the other way was by structuring markets in which people use and exchange those property rights. Those two ways – in this second edition called 'legal approaches' – were compared. How effective was each in achieving the land use which the society wanted? How efficiently did each use scarce economic resources?

When the opportunity was given to revise that first edition, the sole author asked two colleagues to rewrite the book with him. It was decided to widen the scope of the laws considered; not just planning law and laws on property rights, but also the laws protecting the rights of the citizen against the state. And it was decided to compare legal approaches to spatial planning not just in their effectiveness and economic efficiency, but also in their treatment – explicit or implicit – of distributive justice, and in their treatment of legitimacy – does the citizen accept the legitimacy of spatial planning?

This second edition is the result. The first edition has been largely rewritten; the scope is wider and the treatment more systematic. We hope that it will be found useful by those training to be spatial planners, by those practising spatial planning and by those trying to understand and explain spatial planning.

Barrie Needham, Edwin Buitelaar and Thomas Hartmann

Preface to the First Edition (2006)

I have become increasingly convinced of the importance for land-use planning of a good knowledge of law and of economics. With this book I want to show how that knowledge can be used to understand the practice better and also to improve it.

My knowledge of economics began with a university study and has been developing ever since. My knowledge of law is self-taught, much narrower and is not so systematic. I hope that lawyers reading this book will forgive the intrusion of an amateur into their field and the layman's way in which I use legal terms.

I had the good fortune to be taught economics by those who had been taught, at first and second hand, by Pigou. Pigou wrote in the foreword to his classic *Economics of Welfare* (1932, Macmillan: vii):

> The complicated analyses which economists endeavour to carry through are not mere gymnastic. They are instruments for the bettering of human life. The misery and squalor that surround us, the injurious luxury of some wealthy families, the terrible uncertainty overshadowing many families of the poor – these are evils too plain to be ignored. By the knowledge that our science seeks it is possible that they may be restrained.

That is my aim with this book too. I hope it will not be thought presumptuous if I use Pigou's words, however old-fashioned and outdatedly idealistic, to express this.

Barrie Needham
Radboud University, Nijmegen, the Netherlands

1

Why Planning, Law and Economics Matter

The Aim of the Book

Law is indispensable for spatial planning. That is obvious: if there were no legal powers for implementing spatial policy, the latter would be ineffective. There is even a separate branch of law – planning law – for that purpose. It is true that many spatial planners regard that law as a 'necessary evil'. It is, of course, necessary because spatial planning cannot be carried out without it, but it is 'evil' because it can get in the way of creativity and good solutions. Moreover, law is difficult and, therefore, best left to lawyers, according to many spatial planners.

Why, then, a book about planning and law, moreover a book which is written primarily for planners? There are two reasons. The first is that there is more to law in planning than planning law. In particular, the laws about property rights and the laws about the rights of citizens have a great effect on the practice of spatial planning. Many planners know even less about those two branches of law than about planning law, to say nothing about how those three branches interact. The second reason has to do with how law is applied in spatial planning. Usually an application is chosen which should make the chosen planning policy effective; the wish is that the aims of that policy be realised. That is, of course, good practice. But it takes no account of the fact that the way in which the law is applied can have other effects too. Those considered here are the effects of the spatial planning on economic welfare, the effects on justice for the citizen and the effects on the legitimacy of the spatial planning itself. When choosing how to apply law in spatial planning (and also when considering what would be *good* laws for spatial planning), account should be taken of those other effects too, not just of effectiveness in achieving the adopted aims.

In short, planning and law is about more than planning law, and about more than effectiveness. A broad knowledge of law is important both for understanding the practice of spatial planning and for practising it in a responsible way. This book provides a general framework for understanding how law functions in and around spatial planning.

The book is called *Planning, Law and Economics*, so it is about economics too. Why? There are already many books which subject spatial planning to an economic analysis (see, e.g., Harrison 1977; Heikilla 2000). The answer is that the way in which law is applied can have important economic effects which deserve special attention. Sometimes it is even claimed that the economic effects are the only ones which should count, and that the law should be applied accordingly. Here that claim is studied critically, for the economic effects need to be considered in relation to the other effects mentioned above – effectiveness, justice, legitimacy.

That is theory. But the task is of more than theoretical importance. In order to illustrate its practical importance, the book begins with a story about practice. It is fictional, but not far from fact, as any planner with knowledge of practice will recognise.

A Planning Story

Part One

Imagine the following. Just outside the centre of a large town is a nineteenth-century neighbourhood. It is, in planning terms, rundown. There are a few hundred small houses, mostly terraced and occupied by older people who have lived there for years. Some of those people own their houses, some rent. There are a few shops and industrial premises. Most of the buildings are poorly maintained, some severely dilapidated, a few even derelict. A property firm is buying houses, dividing them into small flats and studios, and letting them to students and one-person households. And also, it is rumoured, to prostitutes and illegal immigrants. There is no doubt that there is drug dealing on some streets. The shops are marginal and becoming vacant. The industrial premises are used for small businesses such as local builders, car repairs, furniture making, and printing. During the day, the streets are lined with cars parked for free, mainly by people who work in the town centre. And some streets are full of traffic taking short cuts to avoid congestion elsewhere. A pressure group of local residents is agitating for improvements, the local newspaper carries alarming articles and the local politicians are fully aware of the situation.

The local government decides to investigate. It sets up a working group consisting of local politicians, local government officials (employees of the local government with the necessary expertise, including planners), and neighbourhood representatives (people who were active in the pressure group). Anyone can make their opinions and wishes known to this group. It is to report to the local council (the elected representatives) and the report is to be public.

Its findings about the physical and social situation in the neighbourhood can be summarised as follows. Most of the housing is suitable for single-family use and could be improved fairly cheaply. There is a concentration (site 'A') of derelict industrial premises: if those were demolished, the vacant land could be used for a small park, for there is a shortage of public open space in the area. There is no social housing for rent in the area: perhaps a housing association could be found which would provide that on a suitable location. There is another cluster of industrial premises (site 'B'): the buildings are underused but not derelict. There are some economically viable small firms still operating in the area, some of which are looking for better premises. A number of shops are vacant, and that is likely to increase. It would be good for the area if some other use could be found for those buildings besides housing, for a mixture of uses would make the neighbourhood livelier. It was noted that there seemed to be a demand in the town for cheap premises for 'creative enterprises'; perhaps they could occupy some of the vacant shops. And it was clear that the traffic congestion could be tackled only as part of a traffic circulation plan for the whole town.

Other findings of the working group were about the wishes of the local residents. Most of them wanted to continue living there, but in better housing, and without higher housing costs. They wanted less traffic on the streets, easier parking for their own cars, more open space, and they wanted to feel safer on 'their own streets'.

It was not part of the remit of the working group to make a plan for the area, not even to make suggestions about the content of such a plan. Its task was to investigate the possibilities. It interpreted this widely: not just the physical and social possibilities, also the legal possibilities, the possible ways of realising the changes. The neighbourhood representatives and the local politicians wanted to know more about those possibilities, in order to prepare themselves for the discussions which would follow the publication of the draft plan.

The discussions about the legal possibilities are now reported: "Must a plan be made?" was the first question. This could be a formal land-use plan, showing the land uses which the local government wants for the neighbourhood. The significance of such a plan, it was explained, is that if anyone applies to change the land use on a particular location, and if the application does not conform to that plan, it must be, or can be, refused. A plan can prevent undesirable changes. "But", the local officials warned, "making such a plan might not be effective in realising the policy aims. It would work only if people actually apply for planning permission to develop in accordance with the plan. If no one thinks that the neighbourhood has an attractive future, no planning applications will be submitted."

Moreover, suppose that people did, nevertheless, want to apply for planning permission in accordance with the plan. Suppose it is desired that the housing be available to low-income families, as the local representatives wanted, but that someone applies for planning permission to build high-cost housing. "Even if there were a formal plan showing the desired land use, it would not be permissible to refuse that application," the officials said. "Our planning legislation does not allow a distinction to be made between houses of different prices."

So, making a formal land-use plan might not be effective in all respects. As a result, the working group discussed another possibility. This is that the local government starts to acquire properties, improve them, and dispose of them for uses in accordance with the desired land uses. "So, the local government can refuse to sell to speculative property developers?" asked the neighbourhood representatives. Then it was explained to them that a government body handling in property is bound by restrictions additional to those binding a private legal person. In particular, a government body is required to act impartially, which means that it should not discriminate unreasonably between possible buyers.

Suppose – the discussion continued – that the local government acts in the property market within the existing laws. That could be extremely effective in realising its planning aims: but only if the prices were right. High acquisition prices would persuade most owners to sell, and low disposal prices would attract the desired users. But it could be very expensive: the costs would be high and the returns might be low. This consideration did not concern the neighbourhood representatives, who assumed that their wishes were so important that they should be subsidised if necessary. But the local politicians were very concerned. There were fixed budgets for this sort of project, and more used for this project meant less for projects elsewhere. They wanted to use the available resources efficiently.

Perhaps the costs could be kept low if the local government bought the properties anonymously, through confidential intermediaries (as, it was reported, the property firm did when buying houses in the neighbourhood, so as not to arouse suspicion). "But", it was again asked, "is that a legitimate way for a government body to work? Should that not act openly? Can the result, however desirable, justify such a way of working? And how would people react when they discovered that they had been misled?"

Also, the acquisition costs would increase when it became clear that someone – no matter who – was systematically buying properties in the area. Some owners would 'hold out' for very high prices. They are legally entitled to do that, for they hold the rights of full ownership: it is for them to decide whether or not to sell. "If the property owner should refuse to sell for a reasonable market price, then surely the local government can use its powers to acquire compulsorily?" the local politicians asked. The local officials warned that the judge did not always approve applications to purchase compulsorily: that depended on how the relevant legislation was interpreted. Moreover, here too there might be questions of political legitimacy. Expropriation is the biggest possible infringement of property rights, and it should be – in the opinion of many – used only in the last resort. Can its use in this way in this neighbourhood be justified?

Such considerations apply to the *costs* of realising the plan by buying, selling, and improving properties. The *returns* would be the income from selling and / or renting the improved properties. The size of those returns depends upon the details of the desired land use. If it is the intention that the existing tenants be able to move back into their improved housing, the new rentals must not be much higher than the old rentals, so the returns will be low. If (some of) the existing residents would not want to return, should the housing be improved to a higher standard and sold for higher prices? That would change the character of the neighbourhood and reduce the number of cheaper houses available. Could the provision of social housing – there is none in the neighbourhood – compensate for this satisfactorily? And what about 'creative enterprises' in the old shops? They can afford only low rents. The topic of disposal prices raises questions not only about the financial efficiency of this approach, also about justice. Who should benefit from the improvements? The working group came back to this topic.

Then someone in the working group asked the following question. Just establishing a formal land-use plan might not be effective. If the local government were to try to realise the plan by buying and selling properties, that might be effective but not (financially) efficient. Could not both effectiveness and efficiency be realised if the local government invited one or two trusted property developers to do all the buying, selling and improvements, in close cooperation with the local government? Property developers are better informed about supply, demand and prices than the local government. Moreover, they would bear all the financial risks.

All sorts of objections were raised against this. How would the local government choose the property developers? And should the local residents not be involved in that choice? Who will represent the local interests?

If property developers are involved at an early stage, will they not try to influence the content of the plan, so as to ensure a good profit for themselves? Property developers do not want the details of their commercial transactions to be made public, so they would want the local government not to publish financial information about the redevelopment. And local government should be transparent.

Then someone in the working group said: "If the local government tries to realise the plan by buying and selling properties, or by getting others to do that, is it necessary to go to the time and trouble (including all the prescribed legal proceedings) of making a formal land-use plan?" That suggestion was rejected for the following reasons. The local government should be accountable for its actions: it must be able to justify to the citizens its actions on the property market. But – was the response – that does not need a formal land-use plan. Surely it can be done with a proposal (an informal plan) approved by the local council and made public. True, but then consider the following. If the local government (or its agents) starts to buy properties in that location, that will stimulate financial interest in redeveloping the land there, so people will start to submit planning applications. If those are not in accordance with the land use which the local government wants, they cannot be refused without an approved and formal land-use plan showing that desired land use. A final reason for rejecting the suggestion was also a legal one. If the local government should want to use its compulsory purchase powers or other instruments of land management (preemption rights, urban land readjustment, etc.), the legislation might require that the use of those powers be justified as being necessary to realise a formal land-use plan.

The working group made an assessment of its conclusions so far. The local government should make a formal land-use plan, but that on its own might not be effective in realising the desired land uses. If the local government were to acquire (directly or indirectly) property rights in the neighbourhood, probably that would be effective. But would it be efficient? There are two aspects to that latter question. Could the aims for the neighbourhood not be realised more cheaply? You can crack a nut effectively with a big hammer, but equally effectively and much more easily with a small hammer! The second aspect is: could the resources which the local government would have to put into realising the plan – not just money for acquiring and improving properties, but the personnel costs also – not be used better (i.e. more efficiently) elsewhere? At this, the local representatives became very defensive. They wanted the money to be spent in *their* area. "How can anyone determine what 'the best use' of those resources is?" they asked. "That is a political decision. And we are determined to convince the local council to spend the money *here*."

The working group had discovered also that the decision about which legal approach to take is not just about effectiveness and efficiency, but also about legitimacy. Narrowly: does the local government use its powers within the letter of the law? More broadly: does the local government use its powers within the spirit with which the legislators wanted the law to be imbued? And even more broadly: does the local government enjoy the trust of the citizens?

In the meantime, it had become apparent to the working group that it needed more information about the financial aspects of realising the plan by buying and selling properties. The local government officials were asked to investigate this more fully, and to report back. Their report was to be, at least initially, confidential within the group.

The report provided information which gave a new turn to the discussions within the working group. For example, the costs of amicably acquiring a house which a property company had divided into two flats would be considerably higher than for a house in its original state. The question was raised: Is it fair to reward the property company by making it possible for them to realise the profits to be made from those conversions? Another example: the report predicted that the later a property was acquired, the more would have to be paid for it, for the bargaining strength of the seller would then be greater. Different prices for the same property. Is that fair? A third example. Acquiring one of the clusters of industrial premises and replacing it with a small park would cost a lot of money. A new park would increase the value of the adjacent housing. And if that housing is not first bought by the local government, the price increase goes to the original owners who had done nothing to earn it. Is that fair? One more example: improving the properties within the neighbourhood is expected to raise the price of properties in adjacent neighbourhoods just outside the plan area. Those property owners outside would benefit from the actions inside, without having done anything to earn it. Again, a question of fairness.

This gave a new dimension to the discussions within the working group: fairness or, more formally, justice. Some of the members of that group thought that they could use their common-sense understanding of justice: but they learned that the issue was more complex.

One interpretation of justice – it was explained – is that, as long as everyone has acquired the relevant properties legally, the plan should not impose changes on the existing distribution of property rights. Applied to the possible changes in this neighbourhood, if properties are to be acquired by local government, that should be at full, current market value. And any local park should be provided where it would function best, the consequences for property values being of no concern for the local government.

"That can't be right," was the reaction of many. "It doesn't seem to be fair." So, another interpretation of justice was offered. If realising the desired land use should redistribute benefits from property, and if that redistribution be considered socially desirable, then it should be accepted; if not, it should be compensated. For example, if the new park, which would increase the value of adjacent housing, was located near to expensive housing, the resulting price rise should be 'creamed off' (if that were legally possible), but not if the park were near to cheap housing. Yet another interpretation of justice when applied to spatial planning is that such planning be used actively in order to redistribute benefits in a socially desirable way. For example, the benefits from using property in a particular area should be redistributed from rich to poor, without any compensation to the rich. The neighbourhood representatives favoured this interpretation, for they considered themselves poor. There could be no question of redistribution from rich to poor within the neighbourhood, for there were no rich there. So, the redistribution should be given in the form of subsidies (paid for out of taxes raised elsewhere), which would make it possible to redevelop the area to a good standard.

All those investigations and discussions took more time than the working party had been given, and questions were raised in the local press about delays. So the local council asked the working party to make an interim report, which would be made public.

The publication of this report aroused a lot of comments, most of them critical, some even fierce. The property company which was acquiring, converting and letting housing warned that it would not sell any of its properties voluntarily, and that attempts to acquire compulsorily would be fought "up to the highest court in the land" if necessary. Some older owner-occupiers wrote a letter jointly saying that they were happy where they were, had no need for improvements, and did not consider moving to rented social housing to be an acceptable alternative, even if it were provided nearby. Nevertheless, they wanted an end to the prostitution and drug dealing, and would welcome a small park as long as it did not require the demolition of any housing. One old man said that he would not under any circumstances move out of his house, and that in the event of a forced eviction he would ensure that the local press be there to photograph it. The local Chamber of Commerce said that the parking provided in the neighbourhood was important for the businesses in the town centre, and that a reduction in the number of parking spaces on the streets would cause financial damage, which should be compensated.

There were also a few constructive comments. The local university pointed out that the neighbourhood was providing an increasing amount of accommodation for students: the university would like to be involved in any redevelopment, with a view to increasing the quality and quantity of the student housing there. A local housing association expressed interest in providing a block of social housing for rent. The local history society pointed to the important part the neighbourhood had played in the growth of the town, and asked that this be respected in the scale and layout of any new development.

But, in spite of those constructive comments, there was indignation in the working group. This was felt most strongly by the neighbourhood representatives. "We know the district best," they said. "We have been working on this interim report for nearly a year – can't we just ignore the comments?"

"No," said the local politicians. "We have to retain the trust of all the citizens as far as possible. That means listening to them seriously." "No," said the local officials. Their reasons were legal and functional. First, people have a [presumptive] right to give their opinion in this [informal] stage of the plan making. "Which people?" asked the neighbourhood representatives. "Even the businesses in the town centre? Moreover, the local government owns the streets and can, therefore, determine how they are used for parking." The answer was in two parts. That the local Chamber of Commerce is a legally recognised 'interested party'. And that a government body, unlike a private legal person, is obliged to exercise its property rights in a socially responsible way. That applies also to how it regulates parking on its own roads.

The second reason given by the local officials was that people have a legal right to object against a land-use plan *after* it has been made known that the local government intends to adopt it formally. If criticisms are ignored *now*, they will probably come back when the proposed plan is made public. And that can delay the redevelopment for years. Moreover, the objections can not only be against the content of the plan, but also against the procedures which have been followed (or not followed). Forewarned is forearmed. If we take no account of these criticisms, we can expect them to return. If we choose not to incorporate them, we need to prepare carefully our arguments for rejecting them.

Third, said the local officials, perhaps the critics have a good point. Can we learn anything from them which would make the plan better? For example, that the need for industrial premises might be too small to justify providing them, in any case, in new buildings.

There was one comment, totally different from the others, which caused amazement and – among some of the working group – incredulity. "The real question", the comment said, "is how to ensure that the neighbourhood makes the biggest contribution to the economic welfare of the town. The aim", it continued, "should be that the resources of land and buildings in the neighbourhood be used in a way which is economically the most efficient. It should not be assumed", it went on, "that either the local residents or the local government is able to achieve that aim. They have vested interests, and limited knowledge." The comment continued: it might be that 'the market', being impersonal, knows better than the residents how the neighbourhood could make the biggest contribution to the welfare of the town. In that case, it might be best to make no plan at all, and to accept what people do voluntarily with their property rights (subject to certain minimum controls, such as for public health and safety).

The neighbourhood representatives found it a ridiculous argument. "It is our neighbourhood," they said. "We should be able to determine its future." The local politicians said, "If we accept this, it would mean the end of spatial planning. And we have wishes for our town which go beyond 'economic welfare', whatever that may mean."

"What should we do with this argument?" the local officials were asked. "Well," was the answer, "there is certainly one point to which we should pay careful and explicit attention. We are making a plan which will determine, or at least strongly influence, how land and buildings in the neighbourhood are used in the next ten to twenty years, perhaps longer. The land in this neighbourhood is important for the economic functioning of the whole town. For example, suppose that a lot of new office space will be required near to the town centre. Or suppose that the economy of the town develops in such a way that there is a demand for high-quality housing near to the centre."

That answer was understood by the other members of the working group. "But", they said, "the point can be met by placing the plan for this area within the context of the strategy for the development of the whole town. Locations elsewhere could be indicated where the possible demand for office space and expensive housing could be satisfied. Moreover, we want the traffic problems in this area to be solved, and that too requires looking outside the area itself."

"That is true," was the answer. "But there is another aspect to this argument. Namely that we, a working group set up by the local government, might not able to say what would be the 'best use' (whatever that may mean) of the land and buildings in this neighbourhood." "And what can we do with that possibility?" the other members asked. "We must acknowledge that our own arguments might be wrong. That is not a reason for not making a plan and trying to realise it. At the same time we should be aware of what are called 'market signals', indicating that there might indeed be other land uses in the neighbourhood which would be better for the town. That requires that there must be enough flexibility in the plan to allow for adaptations." The local representatives were not happy with this: they wanted their own interests to be incorporated in the plan, and in such a way that they could be certain that those interests would be protected.

The local officials then referred to a point in the comments of the Chamber of Commerce which had, until then, received little attention. It was the suggestion that cost–benefit analysis be carried out, comparing the economic costs and benefits of the proposed land-use plan with those of alternative land uses in the neighbourhood. "There is a connection", the officials said, "between this suggestion and the argument about the neighbourhood as an 'economic resource'. Both are directed to achieving that use of land and buildings in the neighbourhood which would be economically the most efficient, that is the use which would contribute most to economic welfare."

That explanation was clear to the other members of the working group, even if the implications were unacceptable to them. "Should we then get someone to make that cost–benefit analysis? But what if it should lead to the recommendation: total redevelopment of the neighbourhood? That would respect neither the present uses nor the present users. And we have learned to think in terms of justice and the distributive effects of spatial planning. Would a cost–benefit analysis take account of those effects?" The working group decided not to recommend that a cost–benefit analysis be carried out, giving as argument that it would delay much-needed improvements even more. Nevertheless, the local officials warned of the possibility that someone might make a legal objection that the local government had not used its powers objectively, rationally, impartially, and so on.

The working party then took stock of the situation.

A lot of information had been collected about the circumstances – physical and social – in the neighbourhood and what the various people with interests there saw as problems; also what those people would like to see done as solutions to those problems. In addition, information and opinions had been collected which were relevant for: predicting the effectiveness and efficiency of measures which could be taken; understanding the possible contribution by the neighbourhood to the economic welfare of the town; how the fairness of possible measures might be judged; and forecasting how people would assess the legitimacy of the local council's plans and policies for the neighbourhood.

The working party submitted to the local council a report of its findings. It included also questions about the more abstract points which had arisen during its discussions, although that was outside its remit. What the council did with this information is reported in part two of this story, to be found in Chapter 9 at the end of this book.

Introducing and Defining the Terms

The topics raised in that planning story are explored in this book in more abstract terms. To do that, the key concepts – those which are used throughout the book – need to be defined and described; also the relationships between them. That is done below.

The Activity of Spatial Planning

There are many different definitions of spatial planning (or land-use planning). Most of them have in common that they refer to 'the future' and 'space' in a wider or narrower sense (Hillier 2010). In addition to these two aspects, spatial planning must always be considered as some sort of state or public intervention. Thus, spatial planning is the activity of a state or public body which has been given powers and responsibilities expressly to act on behalf of the public when it prepares and implements decisions about how land and buildings within its jurisdiction should be used. This organisation can be a municipality, a region, or a county, or a province, even the whole national (or federal) state.

If such a body wants there to be a particular use of land within its jurisdiction, that is in order to achieve certain aims. The content of the aims are chosen as being in the interest of those in its jurisdiction, and they can be limited or far-reaching. The state body might, for example, want to do no more than to create the conditions under which its citizens can act in ways which do not harm others, or to create conditions which give its citizens the opportunities to deploy their initiatives. If that is the choice, the state body imposes the conditions and leaves the outcome (the land use) open. Or the state body might have aims – such as stimulating employment, improving the town centre, better amenities in housing areas, better access to work and housing, nature protection – in order to achieve which a specific land use is desired.

There might, in addition, be aims regarding the distribution of land; that is, who is entitled to the land in question. Sometimes those aims are stated explicitly in the planning law.

In order to achieve those aims, the state body might decide that the land – which can include the buildings on the land, mineral resources, water, infrastructure – should be used (or owned) in a particular way.

The aims, the land use considered to be necessary in order to achieve them and the proposed ways of realising that land use are formulated in the form of an ordinance, a spatial vision, policy, programme or plan. These need to be democratically legitimised, usually by an elected organ which directs that state body. A state body doing that is called here, for convenience, a *planning authority*.

This activity is called spatial planning here, although when is it practised at a local scale it is sometimes called land-use planning. It is carried out, formally, by a planning authority. This is advised by spatial planners, but this book is not about 'the planners' (although it is addressed to them) but about the state bodies which have been given the legal powers and responsibilities to carry out spatial planning (and to whom many planners are answerable).

Spatial Planning and Property Rights

The spatial, or land-use, objectives of the state body usually concern land over which *others* hold property rights. The exercise of property rights by their owners is protected by law. Spatial planning must take account of that. It can try to influence how people choose voluntarily to use their property rights, and it can impose rules on how those rights may be used. In the latter case, the state body must be legitimised to do that. It must follow a rule of law.

The legal powers for imposing rules on how others may use their property rights are given only to state bodies (the state has the monopoly on the legal use of power). It follows that spatial planning, as described here, is necessarily a public, or government, or state, activity.

Note that in this respect spatial planning, as described here, is different from the activity when landowners make and implement plans for how they want their *own* land to be used. That can be on a large scale, such as when a firm owns hundreds of hectares for its productive activities, or when the owner manages a big private estate. They plan the use of their own land, not the land of others.

The Types of Law Relevant for Spatial Planning

From the above, it follows that spatial planning is inextricably bound up with law. There are three types of law relevant for spatial planning.

There is law which gives a state body the legal powers to impose rules on how others may use their property rights in land. It includes planning law in the narrow sense (such as development permissions and how they may be granted with respect to spatial or land-use plans), also laws on expropriation and preemption, also many building regulations and much environmental law. It is law which gives a planning authority legal instruments for its spatial planning. Here it is called, for convenience, *planning law*.

Law can have another function, too: to safeguard the position of the citizen. In this case, there are laws which protect the citizen from a state body misusing, abusing or using arbitrarily its spatial planning powers. Such protection includes giving the citizen the right to try to influence the content of the spatial planning, the right to object to, and appeal against, the imposition of rules and the right to oblige a state body to observe 'due process'. The laws which guarantee such rights are discussed here under the name of *citizens' rights*.

Even if there were no spatial planning, there would still be laws about how people use their land. For the way in which one person uses his/her land can affect how others use their land. Most countries in most times have found it convenient to have laws about such interactions. Such laws have the following form. Particular ways of using land – such as freehold, leasehold, easements – are defined in law; also how they can be owned and transferred. If a (legal) person exercising one of those rights over land and buildings is hindered by another in that exercise, he/she can ask a court of law to restrain the (legal) person causing the hindrance. That restraint can be imposed, if necessary by force, and in most states, only the state may do that. So, it is the state which has to introduce the relevant laws, also the penalties for transgressing them. The laws which guarantee such rights are discussed here under the name of *property rights*.

The way in which spatial planning is practised is governed by those three types of law:

–   Planning law can influence, if necessary by restricting, how people use their property rights.
–   The exercise of planning law has to recognise citizens' rights.
–   If, for the location where spatial planning is being practised, property rights are already in place, account has to be taken of them in one way or another.

Spatial planning is sometimes carried out using two additional types of law. One is when two or more owners, or the state body and one or more owners, enter voluntarily into a contract regarding how the owner is to use his/her land and buildings. The rules for this are given by contract law. However, as the relevant law is not specific to spatial planning (it is used for commercial deals between all sorts of actors about all sorts of goods and services), it is not discussed systematically in this book. The other type of law which might be relevant is that which gives a state body power to give subsidies or to put levies upon certain types of activities, depending on the location. In spatial planning, this type of law is usually applied incidentally rather than systematically, and for that reason receives little attention in this book.

Public Law and Private Law

Private law is about the interactions (the 'traffic') between citizens. It is useful because citizens want to make agreements between each other, need to be able to resolve conflicts between themselves, need to know whether they can trust each other, and so on. If one citizen should transgress one of those private agreements, another citizen whose interests are thereby damaged must be able to seek redress. If there are to be sanctions, these may be imposed only by the court. But the state is not involved if the damaged citizen does not seek redress.

Public law is about the relationship between the state and the citizen (and between state bodies). If the state imposes rules and the citizen does not follow them, the state may seek redress directly.

The relevance of this distinction for this book is that property rights are rules under private law (as is all contract law, too). On the other hand, planning law (including environmental law, traffic law, etc.), also the laws about citizens' rights, are rules under public law. Because spatial planning affects the exercise of property rights, the relevant laws are not just public (such as planning law in the narrow sense) but also private.

The Rules We Make for Using Land

This is the subtitle of this book, and the 'we' refers to the society where the rules are made. This book is about formal, legal, rules; that is, rules which can, if necessary, be enforced in a court of law.

In a liberal democracy, governmental power is separated into three branches: the legislative, the executive and the judicial branch. This is called the *trias politica*. The legislative branch makes the legal rules which all state bodies must follow (and it makes private law rules also); the executive branch applies the (public law) rules when pursuing public policies; the judicial branch can intervene if the rules are not followed. The types of rules distinguished above – planning law, citizens' rights, property rights – are determined by the legislature, usually a parliament in one form or another. Spatial planning is an executive (or administrative) action of a state body, and is expected to follow the rules set by the legislature. If it does not, the judiciary can be called upon to judge the action.

The rules just discussed are laws. Using them, a planning authority may try to influence how people may use their land and buildings; those planning actions, too, are rules (but not laws). It is helpful to distinguish between different types of rules relevant to land use according to the level at which they are made:

- pre-constitutional;
- constitutional level;
- legislative level;
- administrative level;
- civil society level.

(Moroni, forthcoming)

The top level, *pre-constitutional,* is where abstract principles are discussed and developed in order to be able to write a good constitution. This resembles social contract theories such as the 'veil of ignorance' of John Rawls. Other examples are Thomas Hobbes' Leviathan or the ideas of Immanuel Kant on public order via private law (Sandel 2007, 2010). At the *constitutional* level, a more or less stable and real constitution has to be written by real people. A constitution is a framework of meta-rules (i.e. fundamental principles). It has two functions: first, it creates, and establishes power to, public institutions, and second, it constrains that power, both substantively and procedurally (i.e. who makes decisions and how). The third level is the *legislative* level, at which laws and regulations are made within the constraints of the constitution. It is at this level that national planning laws are made, approved and established. The fourth level is the *administrative* level, where laws and regulations are applied by government bodies. This is the level where government rules that regulate spatial development directly are established and enforced by planning authorities. Fifth, and finally, is the level of *civil society,* where persons and groups make decisions and rules about social and economic affairs within the context of and in compliance with the law.

In spatial planning, the rules established at the legislative level can be regarded as providing *legal instruments* which a planning authority at the administrative level can apply in order to take a *measure.* For example, if there is a law that a citizen may not build without a building permit, the planning authority may use that as an instrument to require that someone wanting to build must first apply for a permit: handling the application to build is the measure. If there is a law that the application must be refused if it is not in accordance with a formally approved land-use plan, the planning authority must make a land-use plan; that, too, is a measure. In such ways, a particular rule is usually part of a system of rules.

The Legal Approach Taken to Spatial Planning

The contents of the three types of law named above (planning law, citizens' rights, property rights) are determined at the legislative level. This is usually national, but in some countries there are sub-national states which may pass supplementary legislation, and sometimes laws passed by supra-national bodies (such as the European Union) must be observed.

Acting within that (national or sub-national) legislative framework, there is usually room for the relevant planning authority (at the administrative level) to choose which rules to apply, based on which laws, in order to pursue its spatial planning. For example, it might wish to use expropriation (compulsory purchase); it might wish to acquire land and therefore to own property rights; it might wish to involve property owners closely in drawing up the spatial plan; it might wish to influence others by providing infrastructure in particular locations; or it might wish to use a mixture of rules.

In that way, spatial planning can be practised by putting the relevant laws to use in a wide variety of ways, singly and in combination. The planning authority can choose, for a particular case or set of cases, which laws to apply and how to apply them. This is called here *the legal approach taken to a particular planning case*.

When making that choice, the planning authority is constrained within the rules made available by the legislature (Gerber et al. 2018). Those rules reflect what the legislature considers should be the possible *legal approaches to spatial planning in general*. For example, the legislation for expropriation will specify the sorts of situations in which that rule may be applied, and the legislation for issuing building permits will specify when they may or should be granted and when not.

Varieties of Legal Approaches to Spatial Planning

Various types of legal approach to spatial planning can be identified, among which are the following.

One approach can be called 'passive' or 'permissive' planning. It uses the instruments given by planning law to impose conditions on how land and buildings may be used. Such instruments have this form: if the land use in this location is to be changed, permission must be obtained beforehand, and that permission will be granted only if the new land use satisfies certain conditions. Or this form: if certain productive activities are to be carried out on this location, the emissions of water, fumes and noise must satisfy certain conditions. Such a legal approach is called 'passive' or 'permissive', because the initiative for changing the land use does not come from the planning authority. The latter sets the conditions under which initiatives will be permitted.

Another approach can be called 'active', because the planning authority takes actions to realise the desired land use directly, rather than waiting for others to take the initiative. It can do this by acquiring the land and buildings and either itself changing them, or selling them to others, who will then change them. The planning authority can acquire the land and buildings either amicably – using private law to obtain the property rights – or compulsorily – using the public law instrument of expropriation. If the planning authority itself develops the land, it uses a combination of property rights and contract law. If the authority sells the land to another who is to develop it, the same legal combination is used.

With the first approach, the activities of private legal persons interacting in the market for land and buildings are *regulated* by a state body imposing rules. With the second approach, the state body itself acts in the market for land and buildings: it is a *market partner*. In a third possible approach, the way in which the market works is not taken as a 'given'. Markets work within a structure of rules given by property rights and contract law. If those rules are changed – the market is *structured* – the outcome, which in this case is the land use, will be different. This is, therefore, a way of practising spatial planning. However, because such market rules can be made only at the legislative level, markets can be structured only at that level. A planning authority then takes the approach 'Let the market do its work, within that structure.'

Those three types of approach

– passive planning, regulating markets by land-use plans, etc.
– active planning, by buying and selling in the market for land and buildings
– structuring the markets for land and buildings

can be combined in many ways.

One way is when a planning authority wants to leave private legal persons as free as possible to act within the structure of market rules, but considers that there should be in addition some public law rules which set minimum conditions aimed at reducing the damage which one 'developer' can cause to another (Alfasi & Portugali 2007). Those public law rules can have the form of an 'urban code' or a set of ordinances, regulating such matters as building safety, fire regulations, street widths and parking standards, water discharge. Whereas regulating markets by land-use plans aims to realise a certain land use specified in advance, enforcing an urban code or set of ordinances leaves the outcome (the land use) open.

Another legal approach to spatial planning sets out by trying to *persuade* private legal persons to produce, voluntarily, a desired land use. That persuasion can take the form of exhortation and public campaigns such as city marketing; or the form of providing infrastructure so as to persuade people to live, work or build in certain desired ways and on certain desired locations; or the form of levies on undesired activities (such as road pricing to reduce road use) or subsidies for desired development (such as relocating to regions with high unemployment).

Yet another legal approach is when a planning authority works closely with developers: how the planning authority uses its powers under public and private law is determined partly in agreement with the developers.

The Citizen and the State

The content of the legislation is a political decision, and it will reflect political attitudes about the desired relationship between the citizen and the state. Particularly important are the political attitudes about the freedom of the citizen in choosing how his/her land it to be used, the constraints which the state may make on that choice, and the ways in which the citizen is protected against a state body (such as a planning authority) misusing its powers.

This book aims to be applicable to spatial planning in liberal democracies. The principle of a liberal democracy is that the state exists for the citizen and not vice versa, and that the citizens choose the political bodies which determine (among other things) the laws and how they are applied. These are the political bodies which determine the rules relevant for spatial planning. Accordingly, most of the practical examples and illustrations given in this book come from four liberal democracies – the United States, England and Wales, The Netherlands, and Germany – which cover a wide variety of planning systems. Parts of the book might be applicable also to other types of political order, such as state socialism or state capitalism or traditional societies, although no claims are made about this possibility.

In practice, there are great differences between liberal democracies, and some of those differences are manifest in the rules relevant for spatial planning. The content of property rights and how they may be protected, the extent and detail in which a planning authority can restrict the exercise of property rights, the way in which the citizen can demand to be protected against the state – there is great variety in such matters between liberal democracies. Differences in political attitudes can be reflected in other relevant ways too, informally. This can be seen in the expectations which citizens have of their planning authority, in the degree to which citizens trust politicians and their officials and in the willingness of citizens to be involved in the processes of spatial planning.

Evaluating the Legal Approach

In a liberal democracy, all types of public policy – including, therefore, spatial planning – can be expected to satisfy certain conditions. These are:

– The policy should be effective and efficient in realising its aims.
– The policy should use scarce economic resources well.
– The policy must be seen to be just, or fair.
– The policy must be seen to be legitimate.

(Salamon 2000; Gerber et al. 2018)

These can be seen as the *external goals* of spatial planning, external in the sense that they are common to all public policy. They are different from the *internal* goals of spatial planning, set by those who pursue it; that is, the goals to achieve which a particular spatial plan or policy is expressly being implemented.

It can happen that one or more of those external goals is deemed to be so important that it is chosen as an aim of spatial planning, or that it affects the choice of the aims. For example, it might be decided that the aim of spatial planning in general – or of a particular practice of spatial planning – is to increase (spatial) justice, or to increase economic efficiency. Or it might be decided that spatial planning should not be practised if it does not satisfy ideas about justice and legitimacy. This is recognised in this book, but the main attention is on evaluating a legal approach taken to spatial planning which is undertaken to achieve other aims (that is, to achieve the internal goals). That evaluation is made by using the four external goals as criteria. The evaluation can be at two levels:

– the level of the (national) legislation (legislative branch)
– the level at which the legislation is put into effect for a particular practice of spatial planning (the executive branch).

Evaluation at the first level is relevant for choices about the relevant legislation (planning law in its wide sense), at the second level for choices about how to use that legislation. In this book it is the latter which receives the most attention.

Policy Effectiveness and Efficiency

Spatial planning is practised in order to realise particular aims in a particular location. The planning should, therefore, be effective in that way. This should be taken into account when choosing a legal approach. That choice should take into account that the effects of taking a particular measure can depend on the particular circumstances. In addition, different planning aims require different degrees of effectiveness, depending on the urgency and priorities. Account should also be taken of the efficiency with which measures achieve their aims; some measures might achieve the aim more quickly, easily and cheaply than others.

Economic Welfare

Spatial planning can have great effects on the way in which economic resources are used and, therefore, great effects for economic welfare. It is not only the use of the land and buildings that spatial planning can affect, but also the resources of travelling time and energy. The legal approach to spatial planning which is chosen will influence, therefore, economic welfare. This is widely recognised and the related arguments derived from welfare economics have often been applied to spatial planning. Because great importance is often given to this consideration, and because there is a strong argument which links economic welfare to legal approaches by analysing the economic effects of legal measures, this consideration is investigated here in some depth.

Justice

Spatial planning can affect the distribution of ownership, use and enjoyment of land and buildings. This is a question of justice: are land and land uses and access to them distributed in a way which is considered to be morally right? There are different, often competing, views on how to approach and answer this question. One approach favours equality, and if spatial planning can reduce the inequality in access to land and the benefits from land, this is just. Another approach accepts some degree of inequality if that is not unfavourable for those in society who are worst off. Yet another approach takes as its starting point how people have acquired their property holdings. If they have done this according to the legal rules, then spatial planning should not change the distribution of those holdings unless by mutual consent. The book will discuss those different views without taking sides.

Legitimacy

The state body which takes measures to realise the planning aims must be able to justify its actions under the rule of law, citizens should be able to challenge planning actions in an independent court of law, and the actions should be seen by (most) citizens as being acceptable. If this is not so, citizens will distrust the actions taken by the planning agencies, and spatial planning will lose political and social support. Moreover, if citizens have the legal right to dispute actions taken by the planning authority, and if they use such rights because they contest the legitimacy of some planning actions, that can cause long delays. Legitimacy is, however, more than legality. It refers also to the requirement that a planning authority should earn the support and respect of the citizens for whom it plans. The legal approach taken should, therefore, be both legal and recognised as legitimate.

Choosing a Legal Approach for the Practice of Spatial Planning

When a legislature determines the contents of property rights, planning law and citizens' rights, it makes political choices (within the structure of the legal system of that country). The criteria on which those choices are based are the four conditions described above. Moreover, how the different conditions are related to each other is also a political choice. For example, if achieving a particular effect is important and urgent (such as flood prevention), then the law might give little attention to citizen participation in determining the policy. And if economic welfare is the main aim, the distributional effects of the chosen policy might be neglected in the law.

Knowing that the extent to which the conditions are satisfied can vary according to the local circumstances, the legislature can allow for that when determining the laws. For example, it might be permitted to expropriate land under some circumstances, not under others.

The same applies, *mutatis mutandis*, when a planning authority decides which of the available laws to apply and how (that is, which legal approach to take). Moreover, a planning authority can decide to combine the planning laws in particular ways, so as to get the perceived advantages of one type while letting the perceived disadvantages be compensated by another type. Even within one plan area, the planning authority can take a variety of legal approaches, one approach for one type of desired land use (e.g. social housing) and another for another (e.g. market housing).

The political considerations behind the choice of a legal approach can have consequences for the choice of the planning aims, for if the only way to achieve a particular aim effectively would require the application of a legal approach which is considered undesirable – for example, it would interfere too greatly with people's property rights, or it would entail state bodies getting too closely involved in market transactions – then the content of the spatial plan will be adapted accordingly.

The Structure of This Book

This book explores certain topics, using this framework. The topics and the chapters in which they are explored are as follows.

The Three Types of Relevant Law

Chapter 2: Property Rights in Land and Buildings
Chapter 3: Planning Law
Chapter 4: Citizens' Rights in Spatial Planning

The Four Criteria for Evaluating the Chosen Legal Approach

Chapter 5: Law and Policy Effectiveness and Efficiency in Spatial Planning
Chapter 6: Law and Economic Welfare in Spatial Planning
Chapter 7: Law and Justice in Spatial Planning
Chapter 8: Law and Legitimacy in Spatial Planning

Synthesis

Chapter 9: Using the Law in Practice
This book aims not only to set out abstract arguments but also to show their relevance and significance for practice. In order to do the latter, Chapters 2 to 8 contain many practical examples and illustrations. In addition, the book begins with part one of a practical (and fictional) planning story, and ends (in Chapter 9) with the second, concluding, part of that story.

References

Alfasi, N., Portugali J., 2007. Planning rules for a self-planned city. *Planning Theory*, 6(2), 164–182

Gerber, J.-D., Hartmann, T., Hengstermann, A. (eds.), 2018. *Instruments of land policy: Dealing with scarcity of land*. Abingdon, UK: Routledge

Harrison, A. J., 1977. *Economics and land use planning*. London: Croom Helm

Heikilla, E. J., 2000. *The economics of planning*. New Brunswick, NJ: Centre for Urban Policy Research

Hillier, J., 2010. Introduction: Planning at yet another crossroads? In Hillier, J., Healey, P. (eds.), *The Ashgate research companion to planning theory. Conceptual challenges for spatial planning* (pp. 1–34). Farnham, Surrey: Ashgate

Moroni, S., forthcoming. Constitutions, laws and practices: Ethics of planning and ethics of planners. In Salet, W. (ed.), *Routledge handbook of institutions and planning in action* (pp. 185–195). London: Routledge

Salamon, L. M., 2000. The new governance and the tools of public action: An introduction. *Fordham Urban Law Journal*, 28(5), 1611–1673

Sandel, M. J., 2007. *Justice: A reader*. Oxford and New York: Oxford University Press

Sandel, M. J., 2010. *Justice: What's the right thing to do?* New York: Farrar, Straus and Giroux

2
Property Rights in Land and Buildings

What This Chapter Is About

One of the ways in which people's legitimate interests in land and buildings are protected is by recognising and protecting those interests as property rights. Property rights are a social relationship between the holder of the right and all others. The content of that relationship, and how it can be enforced (if necessary), is regulated in law. There is a huge variety of property rights, and this can be analysed by identifying some of the dimensions in those rights. Common to all those rights is that they give rights to, and impose obligations on, both the holder of the right and all others. By identifying those rights and obligations, it is possible to distinguish between four types of 'property regime'. The exercise of a property right is never unrestricted and absolute, and it can be limited in many ways, under both private law and public law. Spatial planning is an exercise of public law, and it can restrict the exercise of property rights. For any plot of land, the 'local regime of land rules' – both public and private – can be identified.

Interests in Land and Buildings

People find land and buildings important in their daily life and work. They want to feel secure in their home, knowing that no one may enter without their permission and that no one may evict them summarily. If the house where they live is rented, they want to be sure that the landlord will repair the roof if it leaks; and the landlord wants to know that the tenants will not damage the building or that, if they do, they can be required to pay compensation. If a firm shares parking space with adjacent firms, all those firms want to know how that space is going to be maintained. If a property developer buys land, it wants to know what restrictions there might be on developing it. People living in a street want their neighbours to maintain their houses and front gardens well. And if there is a park in the neighbourhood, they want to be able to use that safely. A farmer does not want people or animals trampling his crops. But he does want access to his land, and that might lie across the land of another. The owner of land might want to pass it on to his/her children when he/she dies, and he/she wants to be certain that this will take place although it might be many years in the future. People enjoy a beautiful landscape, and want to ensure that it remains beautiful so that their children and their children's children can enjoy it too. A firm might have buildings worth a lot of money, and wants to use them as security for raising a loan; then the bank wants to know that the firm's ownership is a secure pledge. In other words: people have *interests* in land and buildings.

Those examples are of people having an interest in land and buildings as they affect him- or herself directly: my house, my firm, my farmland. That interest might extend to the family of the person: our house, our farm, our shop. And it might be an interest which is shared with many others: the street is not mine, but it is important for me as well as for many others; the landscape gives pleasure to me and also to many others. The interest can be even more indirect, but nevertheless very real: A member of a local community might want the town hall and public parks to reflect the pride he or she has in that community. For a citizen of a country it might be important that there are national parks, museums, or ancient buildings which represent to others the status of the country. People might think of future generations too: it is of interest to me how the land and buildings will be when my children, and my children's children, are there to use and enjoy them. The interests a person can have in land and buildings can be much wider than self-interest.

Should Those Interests Be Protected?

Those are interests which a private person, a private firm, a state body, a non-governmental organisation, or any other legal person might have. And it might be that it is in the interest of the wider society that those interests be protected. For example, if people do not feel safe in their homes, there might be social unrest. If a householder cannot pass on his/her house to others, he/she might not maintain it well. If a farmer cannot prevent others from entering his/her field, he/she will produce less food. No one will lend money to a firm if that firm cannot guarantee that it can continue to use its land and buildings productively. In other words, it can be socially and economically important that some interests in land and buildings are protected.

An Interest Which Is Not Protected

If a person's interest is protected, this is necessarily protection against another person. It might be that the limitation on that other person's action is judged to be disproportionately great. Suppose, for example, that someone in the street paints his/her front door bright pink, and that some neighbours find this distasteful: it is against their interests. They might say: "It lowers the tone of the street." They might even say: "It lowers the value of our houses." Should the first person be prevented from choosing a colour for the front door? If it is judged that the restriction would be greater than the damage to the interest, society may not want to protect that interest.

An Interest Protected as a Property Right

If it is decided that the interest should be protected, that can be done by allowing the person with the interest to take action against the person who is threatening or damaging the interest. Society elevates the interest into a property right. A property right is the right to use something in a particular way, whereby the holder of the right may appeal to the law (formal or informal; see below) – that is, to a collective (Bromley 1991: 15) – to protect that exercise against transgressions by others. That 'thing' can be a land or building (immovable property), but also an object such as a motor car, an umbrella or a gun (movable property), and also intellectual property such as a book or a musical composition. In this book, it is only property rights in land and buildings which are considered; rights such as freehold ownership, leasehold, tenancy, easements and ground leases. They are the subject of this chapter.

Property rights are so important for a society that they have been created in one form or another in many societies and for many centuries. Just one example: in the Old Testament (Deut. 27: 17), written in the seventh century BC, it says: "Cursed be he that removeth his neighbour's landmark." The property right of the neighbour is recognised, is established (it is written in stone! – see verse 4) and there is a sanction on transgressors (a curse!). And today, in international law (such as the European Convention on Human Rights) and in most constitutions in liberal democracies, property in land is well protected.

An Interest Protected by Public Regulation

There is another way in which interests in land and buildings can be protected. Suppose that some people do not want a firm to extract minerals in an area with a beautiful landscape and a rich and scarce biotope. The state can say that those people's interests are legitimate and deserve to be protected.

Sometimes this is discussed in terms of 'the public interest': it is in the public interest that the landscape and the biotope are protected. But that term can detract attention from the fact that it is individual persons who have that interest; also from the fact that not all individuals might share that interest. Whatever the terms used, the state can recognise the interests by requiring the firm to apply for permission to extract the minerals. Permission might still be granted – but only after careful consideration of the pros and cons. This type of public intervention in the use of land and buildings is the subject of Chapter 3. (Chapter 4 complements this by discussing the protection of the citizen against the misuse of such public actions.)

The Choice: Protection by Property Rights or by Public Regulation

In some circumstances there can be a realistic choice between those two ways of protecting an interest in land and buildings. If, for example, it is in a homeowner's interest that the neighbour does not use his/her back garden for repairing cars commercially, that interest can be protected either by allowing the house owner to take action against the neighbour for interfering with his/her property right, or by a state body making a land-use plan which prohibits the use of the land there for anything other than housing. In other circumstances, it is clear that an interest can be protected better by property rights (e.g. the complementary interests of the landlord and the tenant) or by public intervention (e.g. the interests of the residents of a town that the town centre be redeveloped). This choice is discussed further in Chapters 5 and 6 (see also Needham 2006).

The choice will take account not only of effectiveness and efficiency, but also of political preference. That the state should create (or recognise) property rights, and if necessary uphold them, is largely undisputed: a system of property rights is essential for the 'traffic' between citizens. "[T]he system of individual property is the most important guarantee of freedom, not only for those who own property, but scarcely less for those who do not" (Hayek 1979 [1944]: 80). Protecting interests in land and buildings by public regulation (such as spatial planning) is always *additional to* protection by property rights. Because protection by spatial planning can conflict with the protection by property rights, there might be political objections to spatial planning.

Property Rights and the Value of the Interest

The decision whether to protect an interest at all, and if so whether to do that by making it a property right, is influenced by the value of the interest to the holder of it. This is so whether the interest can be traded (when it has an exchange value) or not (when it can still have a use value). (The difference between a property right which can be traded and one which cannot is set out later in this chapter.) It can be decided to protect an interest even if the value of that interest is not monetary but nevertheless great, such as the interests of many people in crossing private land (a public right of way). If the value is great, the holder of it will want to protect his/her interest by having it made into a property right which can be defended in law. In that way, creating a property right gives it a recognised value. Much empirical research has shown that when circumstances change so that the value of an interest increases, the law is often changed to make the interest into a property right (e.g. Davis & North 1971).

Protection by Private Law or by Public Law

In Chapter 1 the distinction between private law and public law was made. Private law is sometimes called civil law, and public law sometimes called administrative law. The distinction is nicely described by Ogus (1994: 2): private law is 'facilitative', whereas public law is 'directive'.

Private law sets legal rules for 'the traffic' between legal persons; that is, how citizens and organisations interact with each other. Note that state bodies also are 'legal persons'. They too have to follow the private law rules about contracts, property rights etc. (unless there are separate rules for state bodies). For example, a municipality with a town hall either rents or owns it, using private property rights. It uses private property rights also when it restricts access to the town hall between 08.00 and 19.00 hours on weekdays and closes it on Sundays. When, as often happens in the Netherlands, a municipality acquires land and then disposes of it for building development, it is using private law.

The essence of private law is that the state makes the rules for the private interactions between legal persons, but enforces them only if one of the parties requests it. If someone damages the interests of another in a way which is deemed to be unlawful, and if the damaged person takes no action, then the state does not intervene. If the damaged person does take action, a state body (the courts) will investigate whether a right has been transgressed. And if so, what sanctions should be imposed on the transgressor. Suppose, for example, that a landlord wants to evict a tenant, contrary to the terms of the tenancy agreement. The tenant can claim that the rules between him/her and the landlord have been transgressed. He/she has to appeal to the court to take the necessary action, because he/she, as a private legal person, cannot impose sanctions on another legal person. But the law

comes into action only if a legal person takes the initiative. There is a tri-
partite relationship: between the plaintiff (in this case, the tenant), the
defendant (in this case, the landlord) and the state.

The essence of *public law* is that the state can intervene without being
requested to do so. If a state body has decided that the interests of the
residents in a neighbourhood should be protected by a ban on the use of
land there for industry, and if someone nevertheless uses land there in
that way, the state body may (sometimes it is obliged to) take action
against the transgressor directly; that is, irrespective of whether the residents complain. The relationship is bipartite: between the transgressor
and the state. A state body may also, in order to allow a town to expand,
change the use permitted on peripheral land from agriculture to housing,
without permission from the owner of that land. And if the owner does
not want to change the way his/her land is used, the state body might be
able to acquire it compulsorily.

Property rights are established, demarcated and protected under private
law, whereas planning law falls under public law.

Property Rights as a Social Relationship[1]

Protecting an interest by making it a property right requires that sanctions
can be imposed on someone who transgresses it. The right of one person to
use a particular thing in a particular way implies an obligation on others;
namely, not to interfere in the exercise of that right. (This is discussed more
fully later in this chapter, in terms of 'rights and duties'.) The result is a set
of rules and expectations within which people interact. It is for that reason
that a property right is a relationship between persons about a thing, not
between a person and a thing. (Denman & Prodano 1972: 22). It follows
that, contrary to common usage, 'property' is not land and buildings, but
the rights in them (1972: 20).

The set of relationships can be politically contested, especially if the
obligations are felt to be too heavy or unfairly distributed. Some people
even consider the existence of (certain types of) property rights to be
socially unjust. "Property is theft", said Proudhon in 1840. Considering
that usually a few rich people own a large amount of land and buildings,
they have more rights than poorer people, and poorer people more obligations than rich people. Moreover, the rich can use their power to
acquire even more land and buildings, and thus increase their power over
others. If the poor people own too little land and buildings for their sub-
sistence and the rich more than enough, that too can lead to property
rights being contested.

It is not surprising that the defence of property rights is often associated with the protection of privilege. Nor is it surprising that spatial planning has sometimes been 'captured' by privileged groups in order to protect their property rights (in any case in the United States; see Jacobs & Paulsen 2009).

On the other hand, if the poor and weak have some rights in landed property (which need not be full ownership, as tenancy is also a right), that can protect them from the more powerful: they cannot be summarily evicted from their property. Those against whom people are protected can include state bodies with property rights. If there is no explicit recognition and protection of property rights, the rich and powerful always win over the poor and weak. It is, however, a condition for that protection that the law be applied in the same way to everybody, irrespective of wealth, origin, sex, race.

How Property Rights Are Regulated in Law

Property Rights, Informal and Formal

An interest in real estate is not a property right unless sanctions can be imposed on those transgressing the interest. In a modern society, the state has a monopoly of power over its citizens. The protection of a property right has, therefore, to be regulated by the state. This is called a 'formal' property right.

There might in addition be 'informal' property rights. Here, the sanction is imposed by the community; that is, by social control. Before the rise of states with recognised authority and law courts, this was the only way of enforcing property rights, and today also informal property rights might be effective (Scott 1998). They might be used to resolve quarrels about land use in a traditional village society, for example, or to protect residents against gangs in informal squatter settlements. And they can be very effective when a group of people try to manage a 'common pool resource' (see later in the chapter). There can, indeed, be 'order without law'.[2]

This book pays most attention to formal property rights in a liberal democracy. In a society with a rule of law (see Chapter 4), if the state is to use its powers under private and public law to allow people to protect or further their interests, that must be regulated by law.

The regulation of property rights is in two parts: determining which interests should be protected in law (i.e. determining what should be a property right), and setting up a legal mechanism for enforcing those rights should the state be asked to do so. Here the attention is to the first way; that is, on how property rights are determined.

How Property Rights Are Determined

In a country with a codified-law tradition (see Chapter 3) property rights can be established by *statute law*; that is, the legislature (the parliament or other government body) makes laws explicitly about property rights – about, for example, agricultural tenancies, or commercial leaseholds, or residential tenancies. Another way is by *case law*. This is when a case concerning a property right is brought before the courts, and the case is not covered (or not adequately) by statute law. The court has to make a decision, which might or might not be appealed against. If the decision is unchallenged or upheld, that decision is then valid for all future similar cases. The property right has been legally recognised, then legally established (the latter can include being incorporated into statute law). If the country follows a common-law tradition (see Chapter 3), the laws governing property rights are built up entirely by case law.

In such ways, property rights are *created*: "things are not protected because they are property . . . things that are protected become property" (Bromley 1998: 24).[3] It follows that property rights are a social construction – they are created by the society. And they can be changed by society. In this respect, property rights can be regarded as an instrumental variable for policy: they can be changed in order to achieve a policy goal.

When property rights are created or changed, there is usually an underlying political philosophy. Norton and Bieri (2014), for example, point to the difference between Germany and the US with regard to the relationship between the individual and the community, and how this affects the regulation of property rights. In Germany, ownership rights are guaranteed, but the owner is expected to exercise them in accordance with social obligations: in the US, the individual may exercise property rights without restrictions, unless that might harm other individuals.

Changing Property Rights Through Statute Law

Property rights are not changed easily. One reason is that they reflect the political philosophy of the society (see above). Another is that the emotional and commercial value attached to many property rights is so great that there is often political resistance to radical change. Also, it is often the case that the people who benefit from the existing property rights are those with the political power to change, or defend, those rights. As North (1990: 101) emphasises, the direction of institutional change (which includes changes in property rights) is determined by the "relative bargaining strength of the participants". And see Libecap (1989: 22): "Institutional change to promote rational resource use and economic growth cannot be taken for granted. Distributional conflicts . . . can block or critically constrain the institutions that can be adopted."

Nevertheless, public policy often does change property rights, assigning and reassigning them. Sometimes that is done directly, by changing the law itself. A good example comes from Britain in the 1930s, when thousands of ramblers trespassed in a coordinated way on private grouse moors. They put forward the moral argument that a few rich owners of huge landed estates were preventing many ordinary folk from enjoying a harmless walk in the countryside. The trespassers won their case and the law was changed: the unlimited right of the landowners to restrict access to walkers was taken away. If it is felt that the legal position of the landlord with respect to his/her tenant is too strong, the property rights might be changed; for example, reducing the right of a landlord to evict a tenant. If the distribution of property rights over land is generally felt to be grossly unfair, the state might undertake land reforms; for example, breaking up big private estates and allocating the land to small farmers.

More often, property rights are changed indirectly, by creating other laws which allow limitations to be placed on the use of property rights. Spatial planning laws are often of this type. For example, when a land-use plan is legally approved, on the basis of and within the limits of a planning law, it changes the content of the property rights on specific parcels of land. Suppose that a parcel was previously designated as 'industry'. This means that, by a change in use, all uses except industry are forbidden. Then the designation is changed to 'commercial'. A partial right on that land (see below), namely the right to develop, has been changed.

Assigning and Reassigning Property Rights

Informal rights can be incorporated into formal law. Squatting, the occupation of buildings against the wish of the owner, has now in many countries been legally recognised under certain conditions. Sometimes the court will recognise a *customary* right: land owned by one person has been used informally by another person for so long and the formal owner has not challenged it, so the informal use is now registered and protected in law. If a superior owner has 'slept on his rights' (that is, was oblivious of, or neglected, them) and someone has used or occupied that property efficiently for many years, the latter might be entitled to 'adverse possession':

the right of the superior owner is terminated. Sometimes people have acted as though they had a right to do so, although that had not been formalised: it was a *presumptive* right. The law can be changed to give them that right formally, or to make it clear that there is no right. For example, there was a presumptive right to discharge industrial effluents into rivers, until increasing pollution of rivers led to that right being withdrawn. In countries which have been colonised within the last few centuries and where the indigenous peoples are still active, those peoples sometimes claim their *indigenous* rights. The coloniser – since then transformed into the legitimate government – did not recognise those rights, and the people want them back in some form or another (Blomley 2004).

The reasons why the state might change property rights can be new ideas about justice and access to opportunities, technical changes such as when natural resources which were previously 'non-property' (see below) are used so unsustainably that they become scarce, population and economic growth leading to urbanisation of land previously in common and unmanaged use.

Stability in Property Rights

Property rights are changing all the time, albeit slowly. If they changed too quickly, the certainty and security which people want from their interests in land would be endangered. Over time, the changes can be great, as illustrated by comparing the situation now with that in feudal times. Then, the way in which land was held (the 'estate' in English land law) determined the social and political position of the holder. For example, a particular way of holding a piece of land might give the right to appoint the parish priest or the local sheriff. Van den Bergh (1988: 19) describes how in those times a law-abiding citizen might be sitting peacefully in his garden, which is suddenly invaded by the local lord and his friends on horseback, hunting foxes. The citizen has no redress in law, for the local lord owns hunting rights over all the land in his feudal domain. Later the citizen kills a wild pig which is rooting in his garden, only to find himself charged with theft, for the local lord has the right to the game also. The citizen with rights to use land on the lord's domain might have the obligation to work on his fields for so many days a year. But then the lord, holding his estate from the king, might have the obligation to serve in the king's army. It is no surprise that in the sixteenth and seventeenth centuries many people emigrated from Europe to the United States, in order to acquire land under conditions which gave them more certainty and fewer restrictions than the feudal laws gave.

Dimensions of Property Rights

A property right is the right to use a particular piece of land and/or buildings in a particular way. There is a huge variety of 'particular ways' which are protected by law. Some of the more important dimensions of property rights are discussed below.

Full and Partial Rights in Landed Property

There can be very many different and non-competing ways of using the same landed property, and many different people can have the right to exercise those different uses over the same land. There is, for example, the right to occupy a building for a certain period. It might be possible also, during that period, that someone else enjoys a separate right to use just a part of that building. On a piece of rural land, there can be the right to farm it, to hunt over it, to gather wood on it, to cross it to get access to adjacent land: and those separate rights can be exercised by different people at the same time. Those are *partial* rights. What are all the possible partial rights? Can anyone own *all* the rights to use a piece of land? What is meant by 'owning the land'? How is that related to all those possible separate rights? There are two different answers to this question.

The Continental legal tradition (which includes both the Napoleonic and the Germanic legal systems – Zweigert & Kötz 1992) starts with the concept of the 'full' or 'absolute' ownership of a thing, or 'dominium' over a thing. There have been attempts to describe what this could mean, by listing the powers which it gives the owner (a summary is given in Needham 2006: 36–9). However, full ownership gives *all* possible rights, even those not yet thought of.

The Anglo-American land-law tradition (Zweigert & Kötz, 1992) is based on the feudal 'doctrine of estates': an estate is an interest in land which can be defined by and defended at law. Originally, the monarch was the ultimate and only absolute proprietor. No one else owned land, but they could hold an interest in it. With this as the basis, ownership of the land arises when one person owns a certain minimum 'bundle of rights'. In the United States, this way of looking at rights in land has persisted, although the absolute proprietor is no longer the monarch but the owner 'in fee simple'.

The relationship between the full right and the partial rights depends on the legal tradition. In the Continental tradition, the person with the right of full ownership might be entitled to split off certain rights, which are then exercised by others: those others do not own those rights, but are authorised by the owner to exercise them. In the Anglo-Saxon tradition (which includes the United States), the starting point is the partial rights, which are owned by the persons who exercise them. Then there is 'an aggregate which is ownership' (Denman 1978: 28) or a 'bundle of rights' (Cooter & Ulen 2004: 77). Under both legal traditions, there are partial rights which are subsumed in fuller rights (they are 'derivative rights'). For example, a sub-lease is subsumed in a lease, which is subsumed in full ownership. So there are superior and inferior rights, with superior and inferior holders of rights.

The Right of Legal Ownership

What does the right of legal ownership of a piece of landed property (not just of a partial right in it) entail? In particular, under the Anglo-Saxon tradition, which partial rights over a thing must you own in order to be regarded as the owner of that thing? This is not just an academic point, for it is relevant for the question: If public regulations (see Chapter 3) take away some of the rights of the owner, when does that go so far that the property has been in effect expropriated? This is important because usually expropriation must be compensated (and see Chapter 4).

Under Roman law,[4] ownership includes:

–   the right to use the thing (*usus*)
–   the right to the fruits of that use (*usus fructus*)
–   the right to dispose of the thing (*abusus*).

The owner can concede to another, or can lose, the right to use and the right to the fruits, and still remain the owner. For as long as the owner has the right of disposal, he/she still has ownership. It is for this reason that, "ownership is not possession, nor is possession ownership" (Denman 1978: 24). However, some countries have laws which modify that principle. If, in England, someone has land on which there are public law restrictions which result in that land having no more 'reasonable beneficial use', the owner can say: "Although I retain the right to dispose of the land, it is a farce to say that I am the owner". And the owner can require a state body to acquire the land. There is a similar doctrine (inverse condemnation) in American law, in Dutch and German law too (see Needham 2006: 152–3).

The Type of Permitted Use

Two sorts of partial rights can be distinguished (Stake 2000):

- The right to use a piece of landed property in a particular way, where that use can co-exist with other uses of the same land. Examples are a fishing right, a right-of-way, a right to 'the fruits'. These are called 'easements' or 'servitudes'. There can also be a right to use the property as security for a loan; and the right to possess the property if the loan is not redeemed.
- The right to use part of a piece of landed property, where all others are excluded from that part. The part can be separated from the rest horizontally (e.g. a field on a farm, a room in a house), or vertically (e.g. mining rights, air rights). These are called 'leases'.

Duration of the Permitted Use

Full ownership is for an indefinite period, as is a freehold interest. Usually, partial rights are for a limited period. Under Anglo-American property law, only freehold interests are in perpetuity: interests for a limited duration are called leasehold interests. (In other legal traditions, some partial rights may have an indefinite duration; an example is the perpetual ground lease in some Dutch cities.) The duration can be specified as a length of time (e.g. a ground lease for ninety-nine years) or, with some types of right, in terms of the life of the holder of the right (a 'life interest'). If the duration is limited, it is specified to whom the right (the future interest) will revert at the end of the term.

There are rules for terminating a partial right before it has expired. Termination can be requested either by the holder of that right (e.g. the tenant) or by the holder of the fuller right (e.g. the landlord). In such cases, the person requesting the termination might have to pay compensation. Sometimes it can be deemed that a right has expired before the end of the term if, for example, the user has neglected it.

The Right to Split off a Partial Right From an Existing Partial Right

It might be permitted under the terms of the partial right that the holder split off a lesser right over the same landed property. For example, a lessee might be allowed to create a sub-lease (sub-letting). Or that might be prohibited. Some countries will not recognise 'excessive decomposition' – splitting off very many derivative rights – because that could cause economic inefficiency or legal uncertainty.

Types of Permitted Owner

Most property rights can be owned and exercised by all legal persons, individual and collective, private and public (a state body). Then the private law rules apply equally to all those persons. For example, a citizen can own or lease an office block, a private company can do so too, a municipal government also. The rights of ownership and leasehold apply to all those persons equally – unless the law specifies otherwise. (The differences between types of property *regimes* – private, state and shared – are discussed later in this chapter: but that is different from types of property *owner*.)

Some property rights, however, can be exercised only by certain named types of body, usually a state body. For example, ownership and use of a nuclear power station might be restricted to a public body. Private persons might not be allowed to own the strip of land around the coast. In the Netherlands, the right to exploit mineral resources is reserved for the state: the owner of the land surface above those resources has no right to them, nor any right to be compensated for the loss of them. And, in some countries, landed property may not be owned by non-nationals.

One of the biggest differences between the regulation of property rights in liberal democratic countries and in socialist or communist countries is in the rules for who may own which types of property. The concept of ownership is the same, but not who may exercise it (Honoré 1961). In particular, socialist or communist countries often want to ensure that the means of production are not owned by (big) private firms, for that can weaken the position of those who *work* with those means of production relative to those who *own* them. So there are rules, such as that movable property (which includes workers' tools) may be owned by anybody, but land, buildings and other capital goods used for production may be owned only by a state body or by a recognised cooperative.

Types of Permitted Transfer

The most important distinction here is between 'real' and 'personal' property rights (between 'rights in rem' and 'rights in persona', using the Roman law terms).

If someone holds a *real* property right, he/she is allowed to transfer that right to someone else, and the content of the right is unchanged. The right can be sold or given away. The transfer can also be by inheritance: the first holder dies and the person named in the testament inherits the right. Examples of real rights are full or freehold ownership, a right of way, a commercial lease (under English law, but not under Dutch law). Transfer does not change the contents of the right. The right 'runs with the land'. If the holder wants to transfer a real property right, it might be legally permissible that he/she attaches to it additional restrictions on how it may be exercised. That can include the obligation not to do certain things on or with the land (a restrictive covenant), also (but not legally recognised in all countries) the obligation actively to do certain things on or with the land (positive covenants). If the property which is transferred is 'encumbered' with a public law restriction (e.g. because it falls within a particular land-use plan), then that restriction also 'runs with the land'. What is transferred is the regulated right. If that were not so, it would be possible to escape the restriction by transferring the property to someone else.

With a *personal* right, the only person allowed to exercise it is the one named in the legal contract. If that person dies, or terminates the right prematurely, or the right expires, then the right ceases to exist. Many residential tenancies are regulated in that way. There can, however, be (limited) exceptions. For example, if the tenancy has been granted to one person, and that person is married or a registered partner, on the death of the named person it might be possible to have the tenancy transferred to the partner. Sometimes a residential tenancy can be transferred to the children of the holder (but not to the grandchildren).

There is another possibility, but it is rare – namely, that the right is inalienable. An example is a public right of way in England: that can be alienated only by an Act of Parliament.

Property Rights Exercised Alone or With Others

Property rights are exercised by a legal person, and in the examples given so far a single legal person exercises a right to the exclusion of all others. However, some property rights can be shared and exercised jointly by more than one legal person. A well-known example is a condominium right: the right to occupy an apartment is for one person, but the right to use the common parts (entrance, staircase, lift, etc.) is for all apartment owners. In the United States, many housing estates are developed as Home Ownership Associations, whereby the association has the right to impose certain user rules and the owner has the obligation to observe them. These are examples of 'real' rights which 'run with the land': if one of the owners transfers his/her rights, the new owner takes over all the rights and obligations.

When a property right is to be used jointly, there have to be rules (formal or informal) for how the holders of that right exercise their rights, for each holder has responsibility towards all other holders; otherwise, the resource (land, buildings) might not be used well.

How property rights can be exercised jointly has received much attention in the study of what are called 'common pool resources'. These are defined by Ostrom as: "a natural or man-made resource system that is sufficiently large as to make it costly (but not impossible) to exclude potential benefi-ciaries from obtaining benefits from its use" (Ostrom 1990: 30). If there are no property rights (which might be informal as well as formal) over the use of such resources, anyone can use them without restrictions and without obligations: then they might become over-used and depleted. It is even, says Bromley (1991: 31), incorrect to speak of property rights over such a resource, because everyone has rights and no one has obligations. It is an 'open-access resource': non-property or *res nullius*, a 'resource in the public domain'. Since Hardin's article from 1968 – "The Tragedy of the Commons" – all common pool resources are often regarded as being 'commons' and in danger because of wastage and depletion (Hardin 1968).

Ostrom (1990) has shown in extensive fieldwork, however, that in many cases the users of a common pool resource organise themselves (the property rights are exercised jointly) so as to ensure that it is used sustain-ably. That applies not only to common *lands*, which fall within the scope of this book, but to many other sorts of common pool resources, such as knowledge, fishing on the high seas, water for irrigation (Ostrom 2009). Then, open-access *non-property* becomes *common* or *shared* property. There are property rights over it, but they are not necessarily the same as when the exercise is exclusively by one holder (Ostrom 2009). And such prop-erty is private, in the sense that access and use are restricted to named persons. This is nicely illustrated by a public notice, displayed in the Minions Area Heritage Project, concerning the common land around Minions on Bodmin Moor in southwest England: "Commons around Minion are private with no formal right of access."[5]

Combinations of Partial Property Rights

Property rights are usually recognised or created by the law maker (the legislature). And if someone could choose one variable within the dimen-sion of full or partial, one variable within the dimension of permitted use, one variable within the dimension of permitted duration, etc., the number of possible combinations would be huge. Some combinations, however, are not logically possible, such as an indefinite duration with a personal property right, or a new partial right with an indefinite duration splitting off from a partial right of limited duration. Nevertheless, there remain a huge number of possible combinations. In practice, one finds some frequently occurring combinations, determined by the type of activity being carried out. Within the category of residential tenancies, for example, there is similarity between different countries, and the same goes for commercial tenancies, farming tenancies, mining leases.

A Classification of Property Rights

Sometimes, in an attempt to classify property rights, a distinction is made between private property rights and public property rights – which Blomley (2004: 7) calls "the ownership model". The idea is that private persons and organisations have property rights which allow them to limit access to their property, and public persons (state bodies) have other property rights, whereby they have to allow the public to use public property such as roads and parks. Sometimes a third type of property right is distinguished; namely, that which can be exercised by a community (e.g. Denman 1978: 102). These distinctions refer to the differences between the *types of legal person* holding the rights and how they will use their rights:

– A private legal person will use the right to pursue its own aims.
– A community will let it be used by its members.
– A state body will appoint officials to manage the property on behalf of the public.

However, this classification by type of legal person with full ownership does not lend itself for further analysis. For example, a state body which owns land and buildings may use them in precisely the same way as does a private legal person (see the example above of municipal offices). A community which lets its members use its land and buildings will stipulate certain binding rules with its members about that use. And it is possible that a private legal person commits itself to manage its land and buildings on behalf of the public.

For such reasons, Bromley (1991: 31) uses the concept of 'property regimes', distinguished according to the rights and obligations associated with the use and ownership of land and buildings, not according to the type of owner. He distinguishes four types (see Table 2.1).

Table 2.1 Types of property regime

PROPERTY REGIME	Rights		Duties	
	Of holder of full ownership	*Of others*	*Of holder of full ownership*	*Of others*
Private property (may be owned by any type of legal person)	To exercise the right within the legal limitations	To expect that the holder of the right remains within the legal obligations	To remain within the legal obligations	Not to interfere with the right of the holder
Public property (e.g. roads, public parks) (may be owned by any type of legal person, but usually by a state body)	To determine the use and access rules	To use the property, and to expect good maintenance	To maintain the property suitably	To observe the use and access rules
Common (or shared) property (may be owned by any sort of collectivity)	There can be many different types of co-holder, each with its own rights	To expect that the co-holders carry out their specified duties	Each co-holder has specific duties	To respect the rules and the rights
Open access resource (= non-property) (owned by no one)	Everyone has the right to use the resource		No one has any duties	

Source: derived from Bromley (1991)

Limitations on the Exercise of Property Rights

Full Ownership of Land Is Always Restricted

Let us suppose that one person has full and absolute ownership over a piece of land. He/she might be inclined to say, "It's my land and I can do with it what I like." But that is not defensible in law. This is made clear in, for example, Dutch legislation:

> Ownership is the most comprehensive right that a person can have over an object. The owner may use the object freely and exclude all others from it, provided that the use does not contradict the rights of others and that account is taken of restrictions arising out of legal regulations and unwritten law.
>
> (Civil Code, Book 5, article 1)

The German Civil Code (Book 3, title 1, para. 903) says something similar: "The owner of a thing can apply it as he wishes and can exclude all others from it, as long as the law or the rights of others do not prevent this." This is echoed internationally in the First Protocol to the European Convention for the Protection of Human Rights and Fundamental Freedoms (article 1):

> Every natural or legal person is entitled to the peaceful enjoyment of his possessions. No one shall be deprived of his possessions except in the public interest and subject to the conditions provided for by law and by the general principles of international law. The preceding provisions shall not, however, in any way impair the right of a State to enforce such laws as it deems necessary to control the use of property in accordance with the general interest or to secure the payment of taxes or other contributions or penalties.

Restrictions Imposed by Public Law

Those legal restrictions on full ownership may be imposed by the state. For example, in some countries it is a criminal offence to refuse to rent property to someone on grounds of race or sex. It is restrictions imposed by *planning law* to which this book gives more attention, restrictions imposed by land-use regulations, environmental laws, traffic laws, etc. For example, in many countries, it is not permitted to build upon or otherwise change the use of land without first having obtained permission. That is a serious restriction on the exercise of one's property rights. The same applies to restrictions on environmental emissions from activities carried out on private property. Such public law restrictions on the exercise of property rights are discussed more fully in Chapter 3.

Restrictions Imposed by Private Law

Legal restrictions on the exercise of full ownership rights may be imposed by the same *private law* which protects property rights. Suppose, for example, that someone has full ownership of a piece of land, and that he/she wants to dig a big hole in it near to the boundary with the neighbour. This endangers the property rights of the neighbour. The neighbour can require the court to prevent that, if the hole would endanger his/her property (he/she has a 'right of support'). For a similar reason, someone might be prevented from erecting a tall fence on his/her own land or from growing trees near to the boundary.

Such limitations on the use of property in order to protect others can be regulated in other (private law) ways too, such as by *nuisance law* (e.g. against noise nuisance), or by *neighbour law*. In Book 5 of the Dutch Civil Code, for example, it says: "An owner is required to roof the buildings on his land in such a way that water does not run off onto someone else's land" (article 52) and "An owner is required to ensure that no water or rubbish on his land comes into the gutter of someone else's land" (article 53).

Suppose that the person with full ownership has sold or delegated some partial rights to another, such as a ground lease or an occupational lease. The person with full ownership has *voluntarily* imposed limitations on himself/herself, for he/she now has legal obligations to the holder of the lesser right. Those limitations are additional to all those already discussed. And the holder of the partial right is subject to the limitations already discussed, and in addition to the limitations which he/she has *voluntarily* entered into with the owner of the full property rights. Moreover, the holder of a right may enter into obligations under contract law, thus limiting further how he/she may exercise the property right.

The Packet of Rules That Influence the Exercise of Property Rights on Landed Property

Much of what has been said above about the dimensions of, and the limitations on using, property rights can be summarised in Table 2.2. Two dimensions are distinguished (exercising the rights and transferring them), as are two sorts of limitations (those under private law and those under public law).

For the enforcement of these rules there are legal instruments (laws) which may be applied *everywhere* within the legal jurisdiction, or within certain types of locations (such as nature reserves, or coastal areas), or

Table 2.2 Types of rules that influence the exercise of property rights

	The exercise *of rights in landed property*	*The* transfer *of rights in landed property*
Under private law	Rights and obligations which can be imposed as a result of a voluntary agreement, e.g. maintaining boundaries, landlord and tenant agreements, easements Obligations which one party can impose on another, e.g. neighbour law, nuisance law	Voluntary agreements, e.g. sales, creating new rights, splitting off partial rights, reversion of a right when it terminates
Under public law	Restrictions imposed under public law, e.g. planning and building restrictions, environmental controls	Imposed transfers, e.g. compulsory purchase, land re-allocation, preemption rights

Source: author.

within the whole country. The planning authority might decide to apply those instruments in the form of planning measures. It might choose to take measures which are the same throughout its jurisdiction, such as local building bylaws or an urban code. The national government might require the same for certain national ordinances and environmental rules. These are *locationally generic* rules. Alternatively, or in addition, the planning authority might decide to give a particular content to those rules according to the location: then they are *locationally specific* (see also Chapter 3). For example, there might be a land-use plan for an area, restricting building above a certain height or restricting the types of permitted land uses.

Taking into account that there are usually specific property rights, entered into voluntarily, which apply to specific plots of land, in addition to any public law rules which affect the exercise of those property rights, in principle it is possible to specify a 'packet' of property rights that apply to a particular plot of land: the 'local regime of land laws' (Geuting & Needham 2012: 37–52) or the 'local user-rights regime' (Buitelaar 2003).

Implications for the Practice of Spatial Planning

Spatial planning is the activity by which a state body takes actions to achieve a desired use of land (see Chapter 1). Those actions necessarily have implications for the exercise of property rights, perhaps also for the ownership of property rights: "Planning cannot escape its relationship to property rights" (Jacobs & Paulsen 2009). Good spatial planning is not possible without a good knowledge of property rights in general, nor a good knowledge of the property rights in place in the locations for which plans are made.

Main Conclusions

1. Property rights affect the way in which people (legal persons) interact when using land and buildings. In that way, property rights affect how land is used.
2. Spatial planning can be regarded as a set of rules imposed by a state body in order to influence how people use land and buildings. Those spatial planning rules are different from the rules governing property rights, but they cannot ignore them, for the court can be requested to uphold property rights. It follows that the effect of spatial planning depends on the interaction between its rules and property rights.
3. As a result, effective spatial planning requires a good knowledge of property rights, both in general and with respect to particular plots of land.

Notes

1 Property rights are also an economic relationship: it is not land and buildings which are traded, but rights in them. There is a branch of economics which analyses economic relationships exclusively in terms of property rights (e.g. Barzel 1997; Cooter & Ulen 2004: chap. 4). A strict application of that approach is not necessary for this book.
2 This is the title of a book by Ellickson (1991) in which he studied how a small, isolated and traditional but modern community in California regulated its own affairs.
3 In this respect, property rights are different from human rights. Human rights are regarded as being so universal and fundamental that they should be protected in law.
4 Roman law has a strong influence on the Continental tradition and, because of its effects on the feudal system, also on the Anglo-American tradition.
5 The question of property rights used by one person, used jointly or absent, is subjected to an economic treatment in Chapter 6.

References

Barzel Y., 1997. *Economic analysis of property rights*, 2nd ed. Cambridge: Cambridge University Press

Bergh, G. C. J. J. van den, 1988. *Eigendom: Grepen uit de geschiedenis van een omstreden begrip*, 2nd ed. Deventer: Kluwer

Blomley, N., 2004. *Unsettling the land: Urban land and the politics of property*. New York: Routledge

Bromley, D. W., 1991. *Environment and economy: property rights and public policy*. Cambridge, MA: Blackwell

Bromley, D. W., 1998. Rousseau's revenge: The demise of the freehold estate. In Jacobs, H. M. (ed.), *Who owns America: Social conflict over property rights* (pp 19–28). Madison: University of Wisconsin Press

Buitelaar, E., 2003. Neither market nor government: comparing the performance of user rights regimes. *Town Planning Review*, 74(3), 315–330

Cooter, R., Ulen, T., 2004, *Law and economics*, 4th ed. Reading MA: Addison-Wesley

Davis, L. E., North, D. C., 1971. *Institutional change and American economic growth.* Cambridge: Cambridge University Press

Denman, D. R., 1978. *The place of property.* Berkhamstead: Geographical Publications Ltd

Denman, D. R., Prodano, S., 1972. *An introduction to proprietary land analysis.* London: George Allen and Unwin

Ellickson, R. C., 1991. *Order without law.* Cambridge, MA: Harvard University Press

Geuting, E., Needham, B., 2012. Exploring the effects of property rights using game simulation. In Hartmann, T., Needham, B., (ed.), *Planning by law and property rights reconsidered* (pp. 37–54). Farnham: Ashgate

Hardin, G., 1968. The tragedy of the commons. *Science*, 162, 1243–1248

Hayek, F. A., 1979 [1944]. *The road to serfdom.* London: Routledge and Kegan Paul

Honoré, A. M., 1961. Ownership. In Guest, A. G., (ed.), *Oxford Essays in Jurisprudence* (pp. 107–147). Oxford: Clarendon Press

Jacobs, H., Paulsen, K., 2009. Property rights: the neglected theme of 20th century American planning. *Journal of the American Planners Association*, 75(2), 134–143

Libecap, G. D., 1989. Distributional issues in contracting property rights. *Journal of Institutional and Theoretical Economic*, 145(1), 6–24

Needham, B., 2006. *Planning, law and economics.* London: Routledge

North, D. C., 1990. *Institutions, institutional change and economic performance.* Cambridge: Cambridge University Press

Norton, R. K., Bieri, D. S., 2014. Planning, law and property rights: A US–European cross-national contemplation. *International Planning Studies* 19(3–4), 379–397

Ogus, A. I., 1994. *Regulation: Legal form and economic theory.* Oxford: Clarendon Press

Ostrom, E., 1990. *Governing the commons.* New York: Cambridge University Press

Ostrom, E., 2009. Design principles of robust property rights institutions: what have we learned? In Ingram, G. K., Hong. Y.-H. (ed.), *Property rights and land policies* (pp. 25–51). Cambridge, MA: Lincoln Institute of Land Policy

Scott, J. C., 1998. *Seeing like a state: How certain schemes to improve the human condition have failed.* London: Yale University Press

Stake, J. E., 2000. Decomposition of property rights. In Bouckaert, B., Geest, G. de (eds.), *Encyclopedia of Law and Economics*, vol. 2, *Civil law and economics* (pp. 32–61). Cheltenham: Edward Elgar

Zweigert, K., Kötz, H., 1992. *Introduction to comparative law*, 2nd ed., trans. from original German T. Weir. Oxford: Clarendon Press

3

Planning Law

What This Chapter Is About

If a planning authority is to be able to realise its spatial plans and policies, it has to be able to impose restrictions on the exercise of rights in land and buildings. Legislation – planning law – gives the authority the legal power to do that. Planning law is made within the context of the legal tradition of the relevant country, the context of other relevant laws and the context of ideas about what kind of spatial order is desired. The scope and the content of planning law can differ greatly between different countries, but that law usually contains rules about the same topics: what can be regulated, who may do that, flexibility and discretion, etc. The planning law of a country has consequences for the legal certainty the citizen can expect from a state body and for the organisation of the planning authority.

Planning Law in Its Narrow and Its Wide Sense

Planning law refers here to the laws which give a state body the powers to impose rules on how others may use their property rights over land and buildings. Those 'others' can be restricted, or have obligations laid upon them, in how they do that. Most developed countries have something called 'planning law' in a narrow sense (e.g. rules for using and developing land). And often that 'planning law' is complemented by other laws which regulate how land may be used, such as building regulations and building codes, environmental laws, laws regulating traffic and parking, laws for expropriation and preemption. This is 'planning law' in the wide sense: all the powers under public law which give a planning authority legal instruments for its spatial planning.

A planning authority may use in addition other types of instruments for that, such as providing information, trying to influence opinions, making subsidies available. Those can be used to persuade people to act voluntarily in accordance with the aims of spatial planning, but those instruments are not imposed. There are some financial instruments which can be, and sometimes are, imposed as an instrument for spatial planning, such as land value taxes (Wenner 2016) and local business taxes used to attract enterprises. And the planning authority may use the property rights on its own land and buildings in that way (see Chapter 2). In this chapter, however, the focus is on the instruments under public law – legal instruments which a state body may use for laying obligations on others.

Rules and Rule-Systems

Planning law is a set of rules made at the legislative level to be applied at the administrative level. At the administrative level, a distinction can be made between land-use rules and rule-systems (Moroni et al., forthcoming). The first concerns specific (categories of) actions. An example of a land-use rule is the following: 'In area X, industrial land uses are not permitted.' A land-use plan or an urban code is a rule-system containing many different individual land-use rules. Rule-systems are often horizontally related to other rule-systems (land-use plans are related to other neighbouring land-use plans, for instance), and vertically to other 'decisional levels' (a land-use plan is made on the basis of, and within, the conditions set in a national planning act).

Some Principles Upon Which Planning Law Can Be Based

What Type of Law for Spatial Planning

A system of property rights is sustained by the state in order to facilitate the 'traffic' between legal persons when they use their land and buildings (see Chapter 2). When people exercise their property rights, an order of land uses will arise even if there is no planning law or spatial planning. That order will be shaped not only by the system of property rights in general but also by the particular circumstances: the characteristics of the owners, the spread of the rights between the owners, the natural physical conditions, the existing man-made land uses, the state of technology, the state of the economy, etc. It is an 'order' in the sense of:

[A] state of affairs in which a multiplicity of elements of various kinds [e.g. different land uses] are so related to each other that we may learn from our acquaintance with some spatial or temporal part of the whole

to form correct expectations concerning the rest, or at least expectations which have a good chance of proving correct.

(Hayek 1982: 35; parentheses added)

Spatial planning, when it is applied, imposes an additional set of rules.
The reason why a state body might want to do that is because it regards
the spatial order resulting from the exercise of property rights alone to be
unsatisfactory in one way or another. There might be social problems,
such as concentrations of bad housing, too little public open space, traffic
holdups and poor accessibility. And there might be missed opportunities,
such as to get a better town centre, to increase biodiversity, to attract
employment. In such cases, the legislature might create planning laws
which enable the planning authority to take (what it considers to be)
appropriate planning measures.

The planning law might enable one or more legal approaches to be
taken – in the terms of Chapter 1, for example, by an urban code, land-use
plans, providing land and infrastructure, collaborating with developers.
And if more than one legal approach is available, the planning authority
has the choice: what kind of rules, additional to property rights, do we
want so that our aims are realised?

Spontaneous Order and Designed Order

Hayek distinguished between two types of order: designed or planned
order (*taxis*) and emerged or spontaneous order (*cosmos*) (e.g. Hayek
1982; Moroni 2007). A planned order is one that is designed from the
outside, exogenously, by a designer (such as an urban planner), whereas a
spontaneous order grows through the individual and free actions within
the order (endogenously) (Hayek 1982: 34–52). These individual actions
are intentional; the aggregate result usually is not. But that does not imply
that the result is *chaos*. The difference between taxis and cosmos is one
between the way the order comes about (design vs emergence), not
between order and chaos (Cozzolino 2017).

Different views on the type of (spatial) order which is desired translate
into different sorts of land-use rules. If the planning aim is that a more or less
determined spatial end-state be realised (e.g. a housing estate according to a
certain urban design, a district centre with specified land uses and circulation pattern), then a detailed land-use plan is the appropriate rule-system. If
the end-state is to be left *open* so as to create many options for spontaneous
action, then an urban code (such as a set of ordinances for building, for
subdivisions, for flood prevention) is the appropriate rule-system.[1]

Those are two sorts of land-use rules, which can be regarded as arche-types, within each of which there can be gradations, and which can be combined in various ways. In their archetypical form they correspond to two theories of public regulation, namely *teleocratic* and *nomocratic* public regulation (Moroni 2010). In the teleocratic approach, "planning is the fundamental, unavoidable central means of (public) land-use regulation . . . via a *plan*, itself . . . a directional set of authoritative rules established with the end of achieving a desired overall state of affairs" (Moroni 2010: 138; emphasis added). The nomocratic approach to regulation is one in which rules are "simple, abstract, and general, purpose-independent, and preva-lently negative: that is, basic and plain rules that refer to general types of situations or actions, not to specific ones . . . and merely prohibit individu-als from certain nuisances, rather than imposing some positive obligation" (Moroni 2010: 146). Hayek (1989) talks of "a spontaneous order produced by the market through people acting within the rules of the law of property, tort and contract".

The rule-system – the type of public regulation chosen – has consequences for the sorts of impositions which a planning authority lays on land owners. A detailed land-use plan specifies the desired end-state and is premised on the rule: only if a potential developer submits an application which con-forms with all those details will permission be given. An urban code is much more open. Permission to develop must still be applied for, but the condi-tions which the application must meet are fewer. Typically they relate to the prohibition of particular negative externalities (as discussed in Chapter 6). The rules enable a land use to arise which is not specified in advance, but which – it is assumed – will be socially and technically acceptable as long as the minimum standards are met (Alfasi & Portugali 2007; Moroni 2007). In doing so, urban codes are *framework-rules* that act as a *filtering* device (i.e. as a sieve) which undesirable externality-producing initiatives do not get through. A detailed land-use plan, on the other hand, consists of *design-rules* that work as *shaping* devices for initiatives so as to make them conform with the desired spatial end-state (Moroni 2015).

A qualification must be added here. A detailed land-use plan specifies the desired end result, but rarely does it impose the positive obligation to realise that. In some countries, such as Germany or Switzerland, it is pos-sible to impose a building obligation on the owner of the land (Gerber et al. 2018) – but that is very rare. Usually, a detailed land-use plan con-tains rules such as: on this plot of land, permission to build will be given only if the developer wants to build housing, but will be refused for all other possible uses. The plan proscribes all uses other than housing. It does not prescribe *de jure* that housing must be built, but the owner is left with the choice: do nothing with the land, sell it amicably or compulsorily or build housing on it. *De facto*, however, it usually implies a prescription, especially if the state body is willing to expropriate in case of non-compliance.

The type of rule-system chosen has implications for the relation between the spatial planning and specific locations and maps. A detailed zoning plan contains *location-specific* rules that are *map-dependent*; a map is needed, as these rules apply to specific demarcated locations only. An urban code contains general rules that apply to all land uses or to categories of land-uses, regardless of their location, or to all categories within a large area such as a local government. These rules are therefore *map-independent* (Needham 2006: 22; Moroni 2007; Moroni et al. forthcoming). Houston (Texas), with its 'code or ordinances' and no zoning plans, comes very close to a city with such an urban code (Siegan 1972; Buitelaar 2009).

The differences can be summarised as shown in Table 3.1.

In a particular country, the legislator can make laws which allow the planning authority to practise only nomocratic regulation of land use (e.g. using urban codes only), or to practise teleocratic regulation of land use (e.g. using a detailed land-use plan), or both according to the choice of the planning authority.

Common Law and Codified Law Traditions

In principle, there are two ways in which legal texts can guide the actions of state bodies and citizens. The first way perceives a legal text merely as "raw material for the communicative process" (Engberg 2002: 378), which needs to be interpreted in concrete situations. The second way presumes a

Table 3.1 Two sorts of land-use rules

	Detailed land-use plan	*Urban code*
Spatial order	Designed (taxis)	Emergent (cosmos)
Spatial end-state	Fixed	Non/open
Theory of public regulation	Teleocratic	Nomocratic
Type of rules	Design-rules	Framework-rules
Relation to initiatives	Shaping	Filtering
Location-specificity of rules	Location-specific	Location-generic
Importance of map(s)	Map-dependent	Map-independent

Source: based on Needham (2006); Moroni (2007, 2010, 2015); Buitelaar and Sorel (2010); Moroni et al. (forthcoming); Cozzolino (2017).

legal text to be an imperative in itself, inducing a normative impact on the regulated subject through its wording. The first way is mainly the Anglo-American interpretation and is called common law; the second way is known as codified law.

Common law developed mainly in the English-speaking world (Buchsbaum 2018). The law emerges out of an incrementally built-up record of court decisions based on actual lawsuits. A judge would – in any new situation that is not recorded before – apply a logical *a priori* understanding, based on some general doctrines that provide a basis for an argument.[2] The inherent notion is that, in common law, law is not made by the legislator, but found by the judges (Buchsbaum 2018). This notion gives courts, through their decisions, some freedom to change the law gradually.

In codified law, rules are established in legislation where situations are regulated beforehand, abstractly, comprehensively and exclusively (Fonk 2010). The role of the judges is to interpret existing rules. For specific situations, in order to comply with the law, one needs to reconstruct the intention of a legislator (Stelmach & Brożek 2011). For this, particular methods of juridical analysis have been developed.[3] In such ways, where it is not clear how the codified law should be applied (because all possible applications can never be included in the legislation), court decisions can provide the necessary clarification. Those court decisions might be taken up into the jurisprudence, and might later be used to change the legislation.

Both traditions – common law or codified law – allow planning law, and the discussion in this chapter is applicable to both. However, the way in which that law is made and applied varies according to the tradition: "In common law countries, judges often have considerable power and responsibility for discretionary adaptation, whereas in codified law countries this responsibility is more a burden of the legislature" (Booth 2016). Courts and judges play an essential role in interpreting the law in the common law tradition (Buchsbaum 2018), whereas with codified planning law, conflicts of interests are to be fought out in the parliament and, possibly, through elections.

Although the two traditions – common law and codified law – are well-known, they are tending to converge, in particular in environmental law and legislation (Stein 2013; van Straalen et al. 2018). This is fostered by legal pluralism and international treaties on environmental issues. The convergence of the two traditions means that both become increasingly relevant for many countries.

In practice, there are considerable differences between countries (regions or states if they are the legislature) in the scope and content of the planning law. The topics which are usually found in that law – in any case in developed, liberal democratic countries (see, for example, European Communities 1999; Gerber et al. 2018) – are discussed below.

The Scope of Planning Law[4]

The Purpose of the Law

This can be stated very generally, such as 'a good spatial land use', as in the present Dutch law on spatial planning; or more specifically, such as 'sustainable development, good living conditions, and protecting and improving the physical environment', as in the forthcoming Dutch law on spatial planning. The German planning law says that planning should achieve a sustainable urban development that balances the social, economic and environmental aspects with the interests of future generations. In addition, planning shall pursue a socially just land use and contribute to a humane environment while also protecting natural resources, shall facilitate an ordered urban design and shall protect and develop the local building culture (German planning law, article 1).

Note that planning law is about how land is *used*, not who *owns* it. The measure which is stipulated for a particular land or building applies irrespective of who owns or uses it. However, if the desired use is different from the existing use, if the existing owner does not want to change the use, and if in addition the planning authority considers that it is in the public interest that the new, desired use be realised (and without too much delay), then ownership rights are obstructing the realisation of the desired land use. Such a situation can arise when, for example, a new road is planned or a town centre is to be redeveloped. It might then be possible for the planning authority to acquire the land compulsorily (expropriation). An alternative action might be that the planning authority lays a preemption right on the land: if the existing owner wants to sell it, he/she must first offer it to the planning authority. Restrictions on ownership and disposal are used also in land readjustment schemes (see, for example, Hong & Needham 2007)

How Land Use Is to Be Regulated

The purpose of most planning law is to make it possible to achieve a desired use of land. But, typically, it is *changes* in land use which are regulated. If the existing land use is not in accordance with the desired land use, the planning law usually does not require the existing owner to make the change; the law comes into effect only when someone wants voluntarily to make a change.

The change which someone wants to make might be to the buildings on the land (e.g. putting a building on previously undeveloped land, changing that building), to the way in which the buildings are used (e.g. from a house to a shop, from a factory to an office, from woodland to arable farming), to the access to the public highway, to environmental emissions (including noise) from the land and buildings. Changes are regulated by requiring someone who wants to change the way he/she uses his/her land, to apply for prior permission for that change, whereby if permission is refused the change is forbidden. Alternatively, it might be that (small) changes do not require prior permission, but that they may be checked later and, if found not to comply with the legal requirements, the responsible person can be required to undo them. Not all changes require prior permission: if that was not so, and the smallest changes were included (a dog kennel in the back garden!), the planning system would be unmanageable and socially unacceptable.

The activities that require permission differ substantially between different countries. The German planning law, for example, has a list of twenty-six activities that require permission by planning authorities – ranging from storing certain materials on the land to the colour of the building materials to be used. Additionally, the shape of buildings and their position on a plot of land are determined by the land-use plan. Some planning laws – such as the Swiss law – are experimenting with prescribing what landowners have to do with the land (i.e. not only restricting changes to land use), using instruments such as building obligations (Gerber et al. 2018).

How the Planning Measures Are to Be Justified

In a liberal democracy, a state body has to follow the rule of law and to justify its actions publicly. That applies to the action of applying one of the legal instruments made available in the planning law, such as granting or refusing planning permission, or expropriation. Such measures may not be taken 'ad hoc', but only as part of a consistent policy. The planning law usually requires that justification be given by a policy document which has been formally approved by a state body.

National building regulations are such a policy document. They lay down the technical specifications which a building must satisfy. So, if someone applies to build or to change a building, the application is tested against the building regulations, and only if it satisfies them is permission granted (subject, of course, to other legal requirements being met, such as land use and environmental conditions). Fire regulations are similar, as are many traffic and environmental regulations.

A land-use plan, or a zoning plan, is another such policy document. This will typically specify the permitted use(s) at every location within a plan area. That might include also specifications about the buildings on the land, such as density, maximum height, orientation – even building materials.

Planning law might require the land-use plan itself to be justified by a plan or policy document at a more general level. For example, it might be required that a planning authority, which wants to make land-use policy for a part of its jurisdiction, justifies why it has chosen that content for that plan area. That justification can be given in the form of a 'structure plan' or a 'vision statement' or some such for the whole of the jurisdiction. In such ways, the rules about applying for planning permission are part of a rule-system.

Which State Body May Use Those Planning Powers?

The planning law, within the conditions of any constitutional law, determines which state body may use the planning powers of granting or refusing planning permission, of expropriating, and of making and adopting the relevant plan documents. Those powers are usually granted to the most *local* authority (e.g. a municipality, county, district). They might also be available to a state body at the regional level, perhaps to the national government as well. And they might be granted to a single-purpose, temporary, authority such as a development agency for a new town, or for an inner-city area, or for a land-readjustment area. The details of the powers which may be used might vary between those different state bodies.

Planning Law and Property Rights

Planning law empowers a state body to restrict the exercise of property rights (and see Chapter 2). This is a fraught relationship (see Needham 2006: chap. 1), in recognition of which planning law often includes rules for protecting the interests of those holding property rights. That protection is given in any case by citizens' rights (Chapter 4): planning law might contain additional safeguards, such as financial compensation for loss of value of property rights (see further below).

Horizontal Coordination

A state body usually pursues policies for many sectors besides land use: policies, for example, for education, for employment, for housing, for water. It is then desirable that the state body coordinates its spatial planning policy with its policies for those other sectors ('horizontal coordination'). It might be considered that a local planning authority, such as a municipality, is small enough to ensure that coordination internally. Alternatively, or in addition, the planning law might prescribe the procedures to be followed in order to improve that coordination. That is one way of improving horizontal coordination. Another way is by consultation with related policy sectors, consultation which the planning law might require as part of the procedures to be followed when making the plan.

In such ways, the rule-system of a spatial plan or policy is related to other rule systems for, for example, water, traffic, nature protection.

Vertical Coordination

The spatial planning pursued by one state body might affect land use in other state bodies at a different administrative level. A regional government might, for example, want a particular area to be protected from urban development in order to conserve the existing natural values; that same area falls in the jurisdiction of a municipal government, which wants some of it to be used for an industrial estate. Or the municipal government might want to allow housing to be built in an area which the national government says should be left open to retain floodwaters. For such cases, the planning law might lay down procedures for achieving the desired 'vertical coordination'. Often those procedures are designed to prevent a 'lower' state body from pursuing spatial planning which contravenes the spatial planning policy of a 'higher' state body: there is some degree of hierarchy in the spatial planning of the country. So, for example, a regional government might be given powers to reject or change parts of the spatial planning of a municipal government. Or a national government might have powers to impose land-use changes on a regional or local government, or to require that the spatial planning of those governments is in conformity with the spatial planning of the national government.

In such ways, the rule system of a spatial plan or policy is related to other spatial planning rule systems, at higher or lower administrative levels.

The Content of Planning Law

Planning law goes further than laying down its purposes, giving the powers for how land use may be regulated, specifying how those regulations must be justified, and determining who may use those powers. A lot of supplementary topics are regulated, some of which are discussed below.

The Required Procedures

The planning law will usually include the procedures which a planning authority must follow when it uses its powers. These include rules for how planning permission has to be applied for, and rules for how the application has to be handled. They will include also rules for how the different types of plan and policy documents are to be prepared, published, and legally adopted (also see Chapter 4 on citizens' rights).

Planning Law and Land Values/Rents

Part of the price which is paid for land is often a 'surplus'; that is, that part which is above the alternative use value of that land. The 'surplus' is called 'land rent' and it arises because there is usually, and unavoidably (i.e. even without spatial planning), imperfect competition on the land market. Spatial planning affects the supply of land for particular uses, therefore the competition between suppliers, and therefore land values.

When spatial planning *increases* land values, that is often regarded as an 'unearned' increase, because an action taken by a state body has increased the wealth of a citizen without that citizen having done anything to earn it; can that be justified? Then measures might be taken to 'cream off' some of the value increase – for example, in order to finance public works which are part of the planning project, or simply to raise tax income. If it is only (part of) the 'surplus' which is removed, it is expected that such measures will not increase the value of land or of the building works on it. When spatial planning *decreases* land values, this gives rise to questions of fairness and, therefore, justice.

Rules for Financial Compensation

The planning law might contain provisions for compensating for financial loss, or for taking away some of the financial gain, or for distributing the losses and gains more evenly over the various land owners. (See Chapter 7 for a discussion of justice when spatial planning redistributes property values.)

When the spatial planning uses expropriation, the situation is clear: the full ownership right has been taken away and, therefore, all the value of the property right. It is generally agreed that this should be fully compensated, although there is much variety in how the amount of 'fair compensation' should be calculated (Alterman 2010).[5] The situation is less clear when spatial planning *decreases* the value of property rights by restricting the possibilities of developing land to a 'higher' (that is, more profitable) use: there is great variety between different countries in how this is compensated, or not.

In the United States this has received much attention, for there are two seemingly contradictory principles. One is that 'police power' is the basis for spatial planning. This refers to the capability of the federal states to regulate behaviour and enforce order within their territory for the betterment of the health, safety, morals and general welfare of their inhabitants. If that reduces property values, that is a side effect which does not need to be compensated: it is comparable to the reduction in the value of a motor car caused by the imposition of speed limits in the interests of public safety. The other principle is that which is incorporated in the Fifth Amendment to the American Constitution: "nor shall private property be taken for public use, without just compensation". The Fifth Amendment is clearly relevant for expropriation: if a state body takes away all property rights, this must be compensated adequately. The complication arises when interpreting what falls under 'taking'. Expropriation – taking all property rights – is clearly 'taking', but is taking away some property rights (such as the right to develop to a higher density) also 'taking'? If so, it should be compensated (Alterman 2010).

In the Netherlands, the legal principle is that, if a *change* in a spatial plan reduces the value of some land (down-zoning), the owner might be eligible to compensation, but a landowner is not entitled to such compensation, whenever an *existing* spatial plan does not allow higher-value development. Spatial planning can reduce property rights in another way too, and this has received much attention in the Netherlands. This is when a spatial plan is changed, allowing a property owner to use his/her land in a different way, and that diminishes the value of someone else's land. For example, someone owned a house with a view of the open countryside, However, a new plan allowed that countryside to be developed as an industrial estate, so the value of the house declines. Until the law was changed recently, the house owner could claim damages from the planning authority, even though it was the person who developed the industrial estate who caused the loss of value. Now, permission to develop can be made conditional on compensating for losses to others (Needham 2014: 123–125).

In England and Wales, it might seem that this issue had been resolved in 1947, when development rights were nationalised. And, in fact, that is the reason why there is no compensation for loss of property rights. There are two exceptions. One is, clearly, expropriation (compulsory purchase). The other is if planning restrictions are so severe that there is no more 'reasonable beneficial use' for the land – then the owner can require the municipality to acquire the property.

The financial issue is less clear when spatial planning *increases* the value of property rights. This can arise when a designation is changed to allow a higher value use: the value of that land increases. Less directly, the value of properties might be raised by changes in designation of adjacent properties (e.g. a park is provided, which increases the value of the houses nearby). Also indirect is when infrastructural works provided publicly increase the value of the land and buildings which become more accessible.

Where planning law, or complementary legislation, includes measures for compensating for such measures, the motive is usually distributive justice (also see Chapter 7). This is the idea that it is not fair that some should gain financially and others lose as a result of spatial planning which is supposed to be for the public good. Moreover, if the public opposition to such 'unfairness' is strong, that might undermine the legitimation of the spatial planning. If the law allows value increases caused by infrastructural works to be taxed (value capturing), there might be an additional motive, namely to contribute to the costs of such works. If the law allows some of the development gains arising from a change in permitted land use to be 'captured', that can sometimes be used to finance associated development, as with the 'article 106 agreements' in English planning law. Where the law allows the introduction of transferable development rights within a certain area, the gains to the owners who are allowed to develop within that area are used to compensate other owners in the area who are not allowed to develop there.

Planning Law, Legal Certainty, and Predictability

The planning authority wants to influence, directly or indirectly, the ways in which others use their land and buildings. Those 'others' want to know what the rules are, both insofar as they affect their own decisions ('what may I do with my land?') and insofar as they affect the decisions of others ('what may my neighbour do with his/her land?'). For the owner of land, the way the planning law is applied should be predictable – the concept of legal certainty is explained further in Chapter 4. In addition, the citizen wants to be able to predict what will happen to the land use round about him or her. That is affected by the planning law not only because of the predictability with which that law is applied, but also because the planning law can regulate land use in more or less detail. The less detail, the less the citizen can predict future land use.

Predictability is important for people's peace of mind, and also for the willingness to invest in land and property. Moreover, the profitability of the investment often depends on the adjacent land uses, so the investor will want to be certain that other relevant land uses will in fact be those as stated in the planning policy. Suppose, for example, that a land-use plan shows land on which housing may be built, and suppose that the adjacent land is shown in the plan as remaining in agricultural use, or that no restrictions are to be placed on the use of that land. A building developer is interested in building on the housing land. But not if he thinks that the adjacent land might be used for waste disposal or some other noxious use, either because the rules of the plan are not imposed, or because there are no rules.

The predictability and certainty a citizen wants from the relevant spatial planning policies is affected by the following three aspects of the planning law described below: degree of detail, discretion and flexibility. The less detail the planning rules give, the more discretion the planning authority may exercise, and the more flexibility there is to change the plans, the less the predictability the spatial planning gives to the citizen.

Degree of Detail

Spatial planning policy can be laid down in more or less detail. It might go no further than an urban code or a set of ordinances. If it wants to specify the desired land use for a particular area, it can do that in general or in detail. For example, land might be zoned for any urban use or for any rural use; for all housing or for affordable housing or for housing for rent or for housing with a maximum sales price; for all commercial uses or for just retail; for all offices or for offices with a maximum plot ratio of 3:1; or the detail that the buildings may not be higher than 10 metres or must be constructed from certain materials. That is in theory possible. In practice, the planning law in force might specify the degree of detail in which the planning authority has to determine the content of its land-use policy. In particular, in some countries there is a list of permissible land-use designations (e.g. in Germany, as above, and the 'use classes' in England and Wales), and planning authorities may not approve legally binding plans which lay down other designations.

Often, a planning authority lays down its policy for a particular area in two stages: a statement (which might be in the form of a spatial plan) showing the general aims for the area, and a land-use plan or zoning plan working that out in more detail. The planning law might even require that procedure. Where the land-use plan has the function of guiding the planning authority in handling planning applications (and, therefore, of informing the citizen about which applications may be granted and which refused), the planning law might specify some of the details which must be included in the plan.

Discretion

This refers to the extent to which a planning authority may legally take actions which are not in accordance with (which 'deviate from'), or which cannot be unambiguously derived from, its approved plan or policy, without first changing that plan or policy (see 'Flexibility' below). The law might allow no discretion, as when a land-use plan is legally binding in all its aspects. Then the planning authority, when handling a planning application, is required to grant that application if it is in accordance with the land-use plan, and to refuse it if it is not: there is no discretion. If the land-use plan is but one 'material consideration' (as in the planning law for England and Wales) which has to be taken into account when handling a planning application, the planning authority has more discretion. Or the planning authority might be allowed to (is given the discretion to) deviate from certain national norms – such as for noise nuisance, or environmental emissions, or traffic safety – if that would, on balance, produce better land use. Sometimes discretion can be built into the plan; for example, that the maximum building height is 10 metres, but that 10 per cent more will be allowed in certain circumstances.

Flexibility

Flexibility, in the sense used here, refers to the ease with which a planning authority can change its formally approved spatial plan. A planning authority makes a spatial plan in order to achieve certain goals, taking account of expected conditions such as population growth, technical change, the policies of other state bodies. If a planning authority changes its goals, or if the conditions change in an unforeseen way, it might want to take measures (in particular, granting or refusing planning permissions) that would be contrary to the existing spatial plan. That might be possible if there is discretion and if the desired action falls within the scope of that discretion. But such a deviation might be legally not possible. Then the planning authority might want to approve a new plan which replaces the old one. The planning law specifies the procedures for making and approving a spatial plan (or for modifying or changing an existing plan), and in that way determines how quickly and easily (that is, the flexibility with which) a planning authority can react to changes. In 2010, for example, in response to the financial crisis of 2008 and its effects on real estate markets, the Dutch government very quickly passed a new law called the 'Crisis and Recovery Act' (*crisis- en herstelwet*). The aim was to make it much quicker and easier to change a legally binding plan, by removing some of the citizens' rights to object to or to appeal against the plan.

Material and Procedural Flexibility

Sometimes predictability and legal certainty are discussed using the terms 'material flexibility' and 'procedural flexibility' (where 'flexibility' has a meaning different from that given above). Material flexibility refers to the opportunities given within a particular spatial plan: how strictly are those described? The designation 'mixed urban uses' gives more material flexibility than the designation 'housing with a maximum height of 10 metres'. Conversely, the first designation gives less predictability than the second. Procedural flexibility refers to the opportunities given by the planning law to all spatial plans. If the planning authority is allowed to ignore its spatial plan when handling a planning application, that gives procedural flexibility, as does the possibility of changing a plan quickly and cheaply. However, high procedural flexibility gives low legal certainty. If low material flexibility leads to rigidity in adapting to new circumstances, that can be compensated by high procedural flexibility. And if high material flexibility leads to low predictability, that might be 'compensated' by low procedural flexibility (Faludi & Hamnett 1978).

Predictability and the Rule of Law

The issue of legal certainty touches upon the inherent fundamentals of the *rule of law*. Simply having legally binding rules is not enough (although a necessity)[6] to guarantee the rule of law. Stability of rules is one of the traditional intrinsic features (see Chapter 4) of the rule of law (Moroni 2007). It is based on the idea that

> rules are to be promulgated, implemented, and revised in a way that enables citizens to have reliable expectations – in general terms and over long periods of time – with regard to the actions of others, and in particular of the state itself.
>
> (Moroni 2007: 148)

In other words, rules and rule-systems that apply to specific cases only, and have to be changed or evaded frequently in order to adapt to new circumstances and to be *effective* in that regard (see Chapter 5), are *ineffective* in adhering to the moral ideal of the rule of law.

Condition-Based and Performance-Based Regulations

For any subject that is regulated, the legislator can decide to determine either the precise procedures to be applied, or the desirable outcome to be achieved by a regulation. As the former implies determining precisely the conditions under which certain legal consequences apply (e.g. a building right is granted), it is called condition-based regulation. The second type of regulation, which defines the outcome or performance of some regulated subject, is called performance-based regulation.

With condition-based regulations, an application to develop must satisfy certain conditions which are stated in advance. They are typically phrased in an 'if . . . then' manner. If an application to develop meets certain specified conditions, permission must be granted; otherwise, it must be refused. For example, if a certain construction does not exceed a certain volume, no building permit is necessary (e.g. there is usually no need for planning permission for a dog kennel); or a petrol station may be permitted only if the runoff water is collected and cleaned in a special drainage system. The material purpose of the regulation is sometimes included in the law, but as an editorial 'aside' without severe consequences attached to it (Fonk 2010). Condition-based regulations enable the planning authority to control the conditions precisely, but the 'performance' of the development (i.e. the material consequences of what is developed) is not specified.

Such conditions are well-known in spatial planning, in the rules governing the granting of planning permissions. They have a wider application, too: they regulate environmental issues, noise regulations (e.g. residential areas may not be located in areas where the noise is above a certain level), air pollution, siting of schools, fire brigades, ambulances (e.g. an ambulance needs to reach every place in a certain district within a certain response time), etc.

With performance-based regulations, the desired performance is stated, and if the proposed development will contribute to that performance, the development is allowed. For example, it might be desired that, in a certain location, land uses arise that will increase employment in the area, but what those land uses should be is not specified. Or that a new housing development should not increase the traffic problems in the area; how the developer is to achieve that is not specified. Performance-based regulations can be recognised by a clear target designed to alleviate certain problems or deficits, often in a certain period of time (Albrecht & Wendler 2009). In other words, performance-based regulations take the material consequences of an action – such as allocating a certain land use – into account. In spatial planning this implies that whether a specific land use at a given location is permissible or not also depends on other land uses and on a changing context.[7]

Which of those two types of regulation is chosen has consequences for the type of administration which is needed to implement the regulations. With condition-based rules, the decision whether to allow or forbid a development can often be made easily. The decision flows automatically out of the conditions:

if the conditions are met, the application must be granted, and if not, then it is not granted (Hartmann & Albrecht 2014). It is work for bureaucrats (in the sense introduced by Max Weber in 1922). With performance-based rules, the aim of a regulation is determined by the legislator or by the planning authority itself, but the way to achieve that aim is left to the executive branch of the state body (e.g. how to achieve sustainable land use, a clean environment, etc.; Durner & Ludwig 2008). For example, the German legislator prescribes a "socially just land use". This can be seen as a performance-based approach that asks the executive branch of the state to achieve this. Within performance-based planning law, planning decisions require a thorough and well-reasoned argumentation. This might require considerable professional insight and judgement, not a role which the Weberian bureaucrat (one who does nothing but apply the rules conscientiously) can fulfil.

This has implications for the mode of governance, as well. The enforcement of condition-based regulation requires participation and stakeholder involvement only to the degree required by the law. On the other hand, performance-based regulations can work best when others – those who will bring about the land-use changes – are involved.

Another consequence of the choice between condition-based and performance-based regulations is that the latter can result in citizens in the same position being treated unequally. Suppose, for example, that the 'performance' to be realised is that traffic volumes should not increase by more than 10 per cent. This can be regulated by imposing conditions, such as on the number of parking spaces in new developments. Someone wanting to build offices can expect that the application will be granted if the application includes those parking spaces. Alternatively, the desired traffic 'performance' can be regulated by allowing offices to be built until the maximum increase in traffic has been reached; thereafter, no permissions will be granted. The first applicants have been treated differently from the later ones.

Main Conclusions

1. Spatial planning can aim to allow people themselves to create an unspecified spatial order, with a minimum set of regulations designed to avoid conflicts; or it can aim to achieve a specified spatial order by setting regulations which steer people's actions towards that order; or it can do a mixture of both depending on the circumstances.

2. Planning law gives the legal instruments to a planning authority for its spatial planning. The legislature, when determining the law, can give instruments for either one or the other of the two approaches described above, or for both, allowing the planning authority to choose.

3. Planning law must regulate a number of standard issues, such as the actions which it regulates, how those actions must be justified, which state body may regulate, and the prescribed procedures. But there can be much variation between countries in how those various issues are resolved.
4. Planning law, and the way it is used, gives more or less legal certainty to the citizens. It affects the extent to which actions of the planning authority are predictable and the extent to which the (land-use) consequences of those actions are predictable.
5. Planning regulations can be 'condition-based' or 'performance-based'. Each has different consequences for the type of administration needed, the mode of governance, and the equal treatment of citizens.

Notes

1 "If we want to create new opportunities open to all, to offer chances of which people can make what use they like, the precise results cannot be foreseen" (Hayek 1944: 79).
2 One example of such a doctrine is the maxim *"sic utero tuo et alienum non laedus"*, meaning that one should not use his or her property to injure another, or the doctrine that nobody should profit from a harm he or she induced.
3 A classical method, developed by Friedrich Carl von Savigny around the eighteenth century and still applied, takes four iterative steps when analysing any legal text (Ifsen 2004).
4 Note that in some countries a legally binding land-use plan is regarded as a 'legal text' and thus a part of the planning law. In this book, the term 'planning law' is used in a more limited way: here the term refers to the legislation which gives all planning authorities the legal powers to practice spatial planning, including the power to adopt a legally binding plan.
5 When spatial planning is pursued by the planning authority acquiring land and buildings amicably – in contrast to compulsorily – the loss of value is compensated by the transaction price.
6 For that reason, planning systems that judge planning applications on their merits and not against a pre-set framework of rules, such as British planning systems, do adhere to the idea of the rule of law (Booth 1996).
7 The term 'performance-based' is derived from the US discussion of the topic (Coglianese, Nash & Olmstead 2002).

References

Albrecht, J., Wendler, W., 2009. Koordinierte Anwendung von Wasserrahmenrichtlinie und Hochwasserrisikomanagementrichtlinie im Kontext des Planungsprozesses. *Natur und Recht*, 31, 608–618

Alfasi, N., Portugali J., 2007. Planning rules for a self-planned city. *Planning Theory*, 6(2), 164–182

Alterman, R., 2010. *Takings international: A comparative perspective on land use regulations and compensation rights*. Chicago: ABA Press

Booth, P., 1996. *Controlling development: Certainty and discretion in Europe, the USA and Hong Kong*. London: UCL Press

Booth, P., 2016. Planning and the rule of law. *Planning Theory & Practice*, 17(3), 344–360

Buchsbaum, P., 2018. The role of judges in using the common law to address climate change. In Straalen, F. van, Hartmann, T., Sheehan, J. (ed.), *Routledge complex real property rights series. Property rights and climate change. Land-use under changing environmental conditions* (pp. 132–146). Abingdon, UK, and New York: Routledge

Buitelaar, E., 2009. Zoning, more than just a tool: Explaining Houston's regulatory practice. *European Planning Studies*, 17(7), 1049–1065

Buitelaar, E., Sorel, N., 2010. Between the rule of law and the quest for control: Legal certainty in the Dutch planning system. *Land Use Policy*, 27(3), 983–989

Coglianese, C., Nash, J., Olmstead, T., 2002. Performance-based regulation: Prospects and limitations in health, safety, and environmental protection. *Administrative Law Review*, 55(4), 705–728

Cozzolino, S., 2017. *The city as action. The dialectic between rules and spontaneity*. Milan: Politecnico di Milano

Durner, W., Ludwig, R., 2008. Paradigmenwechsel in der europäischen Umweltrechtsetzung? *Natur und Recht*, 30, 457–467

Engberg, J., 2002. Legal meaning assumptions: What are the consequences for legal interpretation and legal translation? *International Journal for the Semantics of Law*, 15(4), 375–388

European Communities, 1999. *The EU compendium of spatial planning systems and policies*. Luxembourg: Office for Official Publications of the European Communities

Faludi, A., Hamnett, S., 1978. Planning in onzekerheid: Engelse lessen. *Stedebouw en Volkshuisvesting*, 59(3), 139–148

Fonk, C. F., 2010. Die konditionale Rechtssetzung in der Tradition Otto Mayers: ein antiquiertes Normstruktur- und Gesetzgebungsmodell? *Deutsches Verwaltungsblatt*, 10, 626–633

Gerber, J.-D., Hartmann, T., Hengstermann, A. (ed.) 2018. *Instruments of land policy. Dealing with scarcity of land*. Oxon: Routledge

Hartmann, T., Albrecht, J., 2014. From flood protection to flood risk management: Condition-based and performance-based Regulations in German water law. *Journal of Environmental Law*, 26(2), 243–268

Hayek, F.A., 1944. *The road to serfdom*. London: Routledge

Hayek, F.A., 1982, reprint 2013. *Law, legislation and liberty*. London: Routledge

Hayek, F., 1989. Spontaneous (grown) order and organized (made) order. In Modlovsky, M. (ed.), *Order – with or without design* (pp. 101–123). London: Centre for Research into Communist Economies

Hong, Y.-H., Needham B. (eds.) 2007. *Analysing land readjustment: Economics, law and collective action*. Cambridge, MA: Lincoln Institute of Land Policy

Ifsen, O., 2004. Die "klassische" Methodenlehre bei Savigny. *Ankara Üniversitesi Hukuk Fakültesi Dergisi*, 1, 231–250

Moroni, S., 2007. Planning, liberty and the rule of law. *Planning Theory*, 6(2), 146–163

Moroni, S., 2010. Rethinking the theory and practice of land-use regulation: Towards nomocracy. *Planning Theory*, 9(2), 137–155

Moroni, S., 2015. Complexity and the inherent limits of explanation and prediction: Urban codes for self-organising cities. *Planning Theory*, 14(3), 248–267

Moroni, S., Buitelaar, E., Sorel, N., Cozzolino, S. (forthcoming). Simple rules for complex urban problems. Legal certainty for spatial flexibility. *Journal of Planning Education and Research* (under review)

Needham, B., 2006. *Planning, law and economics*, 1st ed. London: Routledge

Needham, B., 2014. *Dutch land use planning: The principles and the practice*. Farnham: Ashgate

Siegan, B. H., 1972. *Land use without zoning*. Lexington: Heath and Company

Stein, P., 2013. *Roman law in European history*, 12th ed. New York: Cambridge University Press

Stelmach, J., Brożek, B., 2011. *Methods of legal reasoning*. Dordrecht and London: Springer

Straalen, F. van, Hartmann, T., Sheehan, J. (eds.) 2018. *Property rights and climate change: land-use under changing environmental conditions*. Routledge Complex Real Property Rights series. Abingdon, UK, and New York: Routledge

Weber, M., 1922. *Wirtschaft und Gesellschaft*. Tubingen: J. C. B. Mohr

Wenner, F., 2016. Sustainable urban development and land value taxation. The case of Estonia. *Land Use Policy*, 7 (September), 790–800. doi:10.1016/j.landusepol.2016.08.031

4

Citizens' Rights in Spatial Planning

What This Chapter Is About

Citizens have the right to protect themselves against certain actions of a state body, including those actions taken for spatial planning. In a liberal democracy, those rights are based on certain principles about how a state body should act – principles such as impartiality, predictability, proportionality. Those principles can be translated into concrete rights in spatial planning, such as the right to be informed, the right to object and the right to go to appeal. The rights can be established in a number of different ways: in international law and treaties, in the constitution, in statute law, in custom and precedence. The nature of the rights and how they are protected can have important implications for how spatial planning is carried out and what it can realise.

Citizens' Rights in Spatial Planning

Chapter 2 introduced the idea of people having 'interests' in the way that land and buildings – both their own and others – are used. It was said that one of the ways in which those interests can be protected, or furthered, is by the use of planning law (in a broad sense). However, the application of the instruments of planning law can also *damage* the interests which people have in land and buildings. Those interests can include the freedom to exercise one's *own* property rights (those which are not already restricted by private law), such as the freedom to change the use of one's land. And the interests can include enjoying the amenity or convenience resulting from the exercise of property rights by *others*, such as enjoying a local park which might otherwise be used for a shopping centre.

Planning law falls under public law, and the essence of public law is that the state imposes measures on the citizen (see Chapters 2 and 3). It is a one-sided relationship. In order to take account of that imbalance, the state has given powers to the citizens by which they can check the state bodies. Those powers have the nature of rights in the sense of the property rights discussed in Chapter 2: if someone is hindered in the exercise of the right, he/she may ask a court of law to act against the transgressor. In this case, the transgressor would be a state body. The rights are based in law – although they might sometimes be unwritten – and, in that way, they are institutionalised.

Ellis (2000) distinguishes between the following rights as expressed in the British planning process: human rights, political rights, administrative rights, property rights, rights of legal redress. When those rights are applicable to spatial planning, they have been called 'planning rights' (Alexander 2002, 2007). Alexander (2002) distinguishes between procedural and substantive planning rights. Procedural rights are about the procedures the state body should follow before taking the action; substantive rights are about the content of the proposed actions and their expected effects. However, the term 'planning rights' can be misleading, as many of the procedural rights apply to other public sector activities, too. Alternative terms are: 'good governance', 'due process of law', "the right to good administration, 'the right to account-ability'. But those have a coverage which is rather narrow, as they exclude matters such as human rights, which also can be affected by spatial planning.

This book will refer to 'citizens' rights in spatial planning'. These rights can be exercised when a citizen has interests in land and buildings which might be endangered by state actions, directly or indirectly.

The State Actions Against Which a Citizen Can Be Protected

A state body can change land use directly or indirectly. Directly is when that body: builds infrastructure such as roads and railways, services land so that it can be built upon, carries out hydrological works such as draining and damming, manages areas with especial natural values, and so on. Those actions can affect the citizen in many ways: accessibility might be changed, land values increased or decreased, nuisance caused or mitigated, valued landscapes altered. And if the land for the works has to be acquired compulsorily, this too is an imposition, on the existing landowner.

In addition, a state body can, with its rules for others, indirectly affect how land is used. If a (legal) person wants to change the use of land, this might be possible only with prior permission from a state body. That per-mission might be granted conditionally, such as when the permission to extract minerals is granted subject to the condition that the affected area be reinstated after the extraction has been completed. The citizen who wants to change the land use is affected directly. And others might be affected indirectly, through the 'external effects'. For example, if permission is sought to build a shopping centre, that development could affect adjacent uses, traffic flow, turnover of other shopping centres, etc.

The citizen might be affected directly (the citizen is a 'second party' to the 'first party', which is the state body) or indirectly (the citizen is a 'third party'). Then a question arises. Should the citizen be protected in one way or another?

Executive Actions

The actions against which a citizen might want to exercise his/her rights are named above. But citizens' rights also cover the *decision* by a state body to take such actions. Building a motorway is always preceded by a formal decision to do so. A land-use plan might have the legal status that applications to build must be handled in accordance with the plan. Then, an objection against the action of granting (or withholding) an application is too late if that action is in accordance with the plan. If someone wants to object against the possibility of a certain type of development taking place, that objection should be made against the decision to adopt the land-use plan. It is for that reason that rights often cover policy *decisions*, when those decisions have the legal status that they determine or directly influence actions of the state body.[1] Both direct actions and the decisions underlying them are called executive, or administrative, actions.

Legislative Actions

The executive actions discussed above are taken within the national legislation. Decisions about that legislation are called legislative actions. The legislation might be such that citizens' rights in spatial planning are affected. If so, there might be legal ways in which a citizen can try to protect his/her rights, in addition to 'through the ballot box' (i.e. by electing political representatives). The right to apply for 'judicial review' of legislation is described later in this chapter.

Democratic Control by an Elected Body of Politicians

The state body taking such actions is usually controlled by a body of democratically elected politicians – for example, by a municipal council. That body is formally responsible for the actions and decisions, even when those are taken by its officials (civil servants, local government officers, etc.). Is that body, representing the citizens, not sufficient to protect the interests of the citizens? Why should citizens have, in addition, 'citizens' rights'? Those questions are especially relevant when decisions are made concerning policy that is, or aims to be, in the interests of all citizens within the legal jurisdiction, such as the policy expressed in a spatial plan.

One reason for giving the citizen rights, in addition to his/her rights to elect a political representative, is that the representative body cannot control the actions of its officials at the required level of detail. Another reason is that such a body might abuse its considerable powers; for example, if it is controlled by a political majority which ignores the interests of a political minority.

The Citizen and the State

The content of those citizens' rights varies according to the political principles held in the society about the citizen–state relationship. This has already been mentioned in Chapter 1, where it was said that this book is about spatial planning in liberal democracies, and that even within this category there can be great differences between societies in how they interpret those principles. Nevertheless, there are two principles common to all liberal democracies and which underlie the content given to citizens' rights (in spatial planning, among other policy fields). These are the following:

Governments Exist to Serve the People, Not Vice Versa

This is the basic principle of a liberal democracy. It might be necessary, for the good of (some of) the citizens, that the state impose actions on citizens. But the justification must always be whether this is good for the citizens. And if it is good for some, but not for others, should those others be protected or compensated?

The Rule of Law

State bodies must be able to legitimate their actions by reference to the relevant law. That law might allow a wide range of actions and some discretion, and might even allow the body to postpone giving its legitimation (such as when a municipal body withholds financial information about deals with landowners and property developers, in order not to jeopardise ongoing negotiations). Nevertheless, a state body is not allowed to act arbitrarily or *ultra vires* (i.e. outside its powers).

How these two principles affect citizens' rights in spatial planning, as well as some of the variation in that between liberal democracies, is set out below.

The Principles State Actions Should Follow

There are many different ways of classifying such principles. For this book, the following classification is used (and there can be considerable overlap between the separate principles described briefly below).

The Principles State Rules in General Should Follow

This book is about 'the rules we make for using land', and such rules (Moroni 2007: 148) should satisfy four (intrinsic) conditions:

- They must be known and 'knowable'. This is the condition of *publicity*.
- They must be stable, so that people can have reliable expectations about the behaviour of others and of the state: the condition of *stability*.
- They must apply equally to all the people specified in the rules: the condition of *impartiality*.
- They must not apply to situations which occurred before the rule was introduced: the condition of *non-retroactivity*.

The second and third conditions are now discussed.

The Condition of Stability

A state which aims to follow a rule of law should have stable laws:

> rules are to be promulgated, implemented, and revised in a way that enables citizens to have reliable expectations – in general terms and over long periods of time – with regard to the actions of others and in particular of the state itself.
>
> (Moroni 2007: 148)

Those rules concern the predictability of state actions – the citizen should know how he/she stands with respect to the state;[2] that is, citizens should be able to know in advance which rules with which content apply to their actions, what they may do and what is restricted. And if he/she breaks the law, he/she must know the sanctions to which he/she will be subjected. This knowledge should be available to the citizen before he/she takes any action, and the citizen should be able to predict reliably how the state will react to him/her action. A citizen wants the same information about the actions of other citizens which might affect him/her, such as the neighbours.

Of course, the citizen might be more devious. He/she might want to do something with his/her land which is *not* in accordance with the spatial plan. But only if all *other* citizens are obliged to follow the plan. For example, he/she might want to be given permission to build a house in the middle of a protected natural area where building is not allowed, but if others too were able to do that it would spoil his/her pleasure in living there. Or he/she might want to build on his/her land higher than the permitted maximum height, while wanting that others are not allowed to do that. That is, a citizen might want to be able to exploit legal uncertainty, with the certainty that others will not be allowed to do that.

Predictability and legal certainty are conditions for robust decisions on investments in the built environment. Such predictability and certainty apply to all aspects – the direct and indirect, as well as the physical and legal, aspects. This implies that not only the actions on one's own property, but also on adjacent properties, is relevant. If someone considers building a shopping centre, the question is not just whether it is permitted by planning law, but also whether someone else in the vicinity will be allowed to build a competing shopping centre. This can be particularly difficult with properties close to administrative borders, where the predictability in an adjacent administration might be lower.

"The liberty that the law is designed to respect is achieved through the guarantee of certainty that the law affords" (Booth 1996). More generally, the citizen should be able to expect of a state body that it gives legal certainty, that it does what is says it will do, and that those statements are reliable. Note that such predictability is also affected by socioeconomic or environmental changes (see Hartmann & Spit 2018).

This is a very demanding ideal and can never be fully met, for a state body can never foresee every possible action by its citizens now or in the future, and it cannot therefore let it be known in advance how it will react in all circumstances. A state body will, therefore, want to be able to exercise its powers for spatial planning in a flexible way. How planning law can make that possible, and the tensions this gives rise to, are discussed in Chapter 3.

The Condition of Impartiality

It is expected that a state body does not discriminate between its citizens. This means that under the same circumstances all people should be treated the same. This is not to say that spatial planning will not harm or benefit some more than others. If an urban area is to be redeveloped, it is inevitable that some will be affected differently from others (e.g. the property owners in the redevelopment area compared with those outside the area), for each occupies a different location (i.e. the circumstances are different, hence the treatment can be). Moreover, "planning, or rather zoning . . . deliberately sets out to be discriminatory" (Sorensen & Auser 1989). But it is not acceptable that some of the existing property owners are treated differently from other property owners in the same location because of their ethnicity, or because they are good friends of the mayor.

The Principle of Rational Justification

The expectation – which might be explicit or implicit – is that the decision has been taken rationally. It should be clear what the objectives of the action are; it should be clear that the relevant information has been collected and the relevant bodies consulted; the reasoning leading to the decision should be clear, including why that decision has been taken and not others; the possible side effects should be considered, etc. Justifying rationality requires that the body acts transparently.

Principle of Proportionality

This might be included in the principle of rational justification but deserves separate treatment. The principle of proportionality is one of the key principles of governmental action: any action by a state body should be appropriate, necessary and suitable (Ellis 1999).

Appropriate means that the negative impact of any governmental action may not exceed the advantages disproportionally. This implies that constitutional rights may not be violated; also that the rights of individuals and the general public need to be fairly balanced. For example, expropriating all landowners next to an industrial site (e.g. a waste incinerator or a chemical firm) in order to prevent externality claims would be a disproportionate action.

The principle of necessity means that, when choosing an action, the state body needs to explore carefully whether there is no alternative action that is less restrictive. This means that, in any situation, the least restrictive action for achieving a certain policy goal has to be applied. For example, if there are externalities from an industrial site, it is not necessary to close the factories if the problem can be solved with less interventionist technical solutions (such as installing a filter).

Suitability requires that the action in question shall result in the desired outcome. So, there needs to be a logical causal relation between the action and the policy goal. Take the example of a municipality wanting to make the town centre more attractive to shoppers. It might suggest doing that by allowing cars to park right outside the shops, but experience has shown that in big town centres that makes shopping more difficult and less attractive; the proposed action would not be suitable.

Ultra Vires

This is the principle that a state body should not act 'outside its powers' (*ultra vires*). This can have a broader application than 'the rule of law' mentioned above (a state body must not use legal powers which it has not been given). There is, for example, the principle of *détournement de pouvoir*: powers provided for one purpose should not be used to achieve other purposes for which they were not intended. For instance, according to EU regulations land-use restrictions for retail development must be used for spatial objectives only and not for protecting retailers from economic competition with other (new) retailers.

The Right to Be Heard

This is the principle that the citizen be involved, in one way or another, in the way that policy for spatial planning is determined. This right might be restricted to 'interested parties' (see below).

The Public Interest

This is the principle that the proposed action always be in the public interest. Alexander (2002) refers to this concept, which has been widely used in spatial planning, as a potential planning right. Then he points out how difficult it is to apply in practice. He distinguishes between the procedural dimension of the public interest, which is included in the citizens' rights discussed above, and the substantive dimension. It is the latter which is so difficult, partly because it is easily interpreted in a utilitarian way; that is, putting all advantages and disadvantages into an accounting framework. That treatment is discussed in Chapter 6. Another criticism is that the term 'public interest' often detracts attention from the fact that it is individuals who are affected by a plan or project, not an anonymous public (and see Chapters 2 and 6). Campbell and Marshall (2000) propose using the idea of "collective responsibilities in relation to the ways society shapes local environments". Because of such difficulties, this book uses the concept of the public interest very selectively.[3]

Translating Those Principles Into Citizens' Rights

Translating the principles into citizens' rights is a political matter. It is for the legislator to decide whether to pass laws which would make it possible for citizens to protect themselves against transgression of those principles, and in what way and how strongly. This is sometimes discussed in terms of participation; the right to participate in spatial planning. That description is, however, too unclear for a useful discussion. In practice, the following aspects are often laid down in law.

The Types of Executive Actions Relating to Spatial Planning Against Which the Citizen Can Exercise His/Her Rights

These can include the following:

- the decision about a planning application
- the decision about the content of a legally binding plan
- the decision to approve a draft version of such a plan
- the decision to adopt formally such a plan.

The Rights of the Citizen When Such Actions Are Taken

These can include the following:

- the right to be informed that a relevant decision is being prepared (e.g. that it is intended to prepare a plan for the town centre, or that a planning application is being considered)
- the right to submit an opinion about the content of that decision (e.g. wishes about the town centre)
- the right to be informed about what the state body declares to be the intended content (e.g. when a draft plan for the town centre has been made)

- the right to object to that intended content
- the right to be informed about the definite content (e.g. the plan as adopted by the municipal council, or that a planning application is to be refused)
- the right to appeal against the definite content before the plan has become legally binding.

There might be exemptions, situations in which the rights may not be exercised. An example is when the decision has to be made and action taken quickly, because of an emergency (e.g. flooding) or of danger to public safety and security.

How the Citizen Can Exercise Those Rights

This can be done as follows:

- The citizen presents to the planning authority views about the proposed action.
- If the planning authority does not incorporate those views in a way which satisfies the citizen, the citizen presents to an independent court a claim that the planning authority has not acted in accordance with the required procedures and in so doing has damaged his/her interests.
- The court judges whether the planning authority has followed the required procedures, and whether it has followed the principles described above, such as rationality, equal treatment, etc.
- If the court judges that the planning authority has not done that, the executive action can be overturned.
- If the court judges that the planning authority has satisfied the procedures and the principles, the executive action is declared valid and the citizen has lost the case.

Who May Exercise Those Rights?

Someone – some legal person – who would be directly affected by an action, or who would be directly affected if the decision to take that action were adopted, is a 'second party' (the first party is the state body taking the action). This is someone, for example, who applies for planning permission, or with property rights on land or buildings within an area for which a legally binding plan is being prepared. This is clearly an interested party and such a party may exercise his/her citizens' rights.

There can be others with a less direct interest, but who nevertheless are concerned about the use of the land and buildings which are being regulated. It might be the neighbour of the person who applies for a building permit to change his/her house. It might be someone who uses a town centre for which a redevelopment plan is being prepared. It might be someone who is concerned about biodiversity and who sees that this might be endangered in a particular area for which a plan is being made, or for which a decision is being given to widen a road. These are 'third parties'. Should they too be able to exercise citizens' rights in the case in question?

There are no simple answers to the question as to who is a third party and with respect to which rights.[4] Narrow boundaries to who has such rights might exclude people with legitimate concerns. Wide boundaries can open the floodgates and include those whose interests are marginal, thus clogging up the procedures. Moreover, the answer might depend on the stage in the decision-making process.

Which Legal Body Protects the Rights?

For each of the rights, the legislation must establish where the citizen may go for redress if he/she thinks that the right has been transgressed. That body should be independent of the state body against which the appeal is made. If the citizen is not satisfied with the judgement passed by a lower court about the transgression, there is sometimes an ultimate court of appeal, and judicial review might be possible.

Judicial Review

The principle of the *trias politica* is that the power of a state is divided and spread between three types of bodies: the legislative, which determines the laws; the executive, which implements the laws as necessary; the judiciary which oversees the legality of the implementation. Many countries have the possibility of some kind of judicial review, whereby an independent court may examine executive actions if a citizen claims that they are in violation of a higher legal authority. In some countries, judicial review can go further: the judiciary can oversee the legality of laws which have been passed by the legislative body (a 'legislative action'; see above), by testing them against more basic laws such as the constitution.[5] This procedure can be used to protect citizens' rights.

The Legal Authority for Citizens' Rights in Spatial Planning

For citizens' rights to be effective, it must be possible to uphold them in a court of law. So they must be legitimated by some formal and legally adopted document. There are several possible authorisations.

International Treaties

The First Protocol to the European Convention for the Protection of Human Rights and Fundamental Freedoms, and its implications for rights in land and buildings, have been mentioned in Chapter 2.

Member states of the European Union have to subscribe to the Charter of Fundamental Rights of the European Union. Article 41 – the right to good administration – says:

1. Every person has the right to have his or her affairs handled impartially, fairly and within a reasonable time by the institutions and bodies of the Union.
2. This right includes:

    – the right of every person to be heard, before any individual measure which would affect him or her adversely is taken;
    – the right of every person to have access to his or her file, while respecting the legitimate interests of confidentiality and of professional and business secrecy;
    – the obligation of the administration to give reasons for its decisions.

3. Every person has the right to have the Community make good any damage caused by its institutions or by its servants in the performance of their duties, in accordance with the general principles common to the laws of the Member States.
4. Every person may write to the institutions of the Union in one of the languages of the Treaties and must have an answer in the same language.

Many citizens' rights in spatial planning can be derived from this.

The Constitution of the Country

This is, for example, the basis for the extremely powerful principle of due process in the US. The relevant clauses are: "No person . . . shall be deprived of life, liberty, or property without the due process of law" (the Fifth Amendment that applies to the federal government) and "No state shall . . . deprive any person of life, liberty, or property, without due process of law" (the Fourteenth Amendment that applies to the states). These are the origins of the concept of 'due process', which is now used in many other countries too (see, for example, Ramraj 2004).

Codified Law or Common Law

A country where (most of) the law is codified will have statute laws containing the rules for citizens' rights and how they may be exercised. There might be one law which applies to all policy sectors. In the Netherlands, for example, there is a *general law* regulating the procedures which must be followed before a state body takes, under public law, *any* executive action which might affect the citizen (*algemene wet bestuursrecht*). There is another *general law* in the Netherlands, which applies to all policy sectors, regulating access by the public to documents produced by and for public bodies (*wet openbaarheid van bestuur*). In other countries, there might be statute laws specific to the policy sector – in this case specifically for land-use, or spatial, planning.

In a country with a common law system (see Chapter 3), precedence and judicial rulings can lead to similar legal procedures being adopted.

A country might have supplementary legislation or case law which confers rights on its citizens so that they can better protect themselves against a powerful government. In the Netherlands, for example, a state body often buys and sells land using private law agreements. But then the public authority is bound to *more* restrictions than a private legal person is. The public authority must follow *all* the rules of private law, and *more besides* – it is more than just one more actor on the land market.[6] The reason why there are more restrictions on a public legal person than on a private legal person, when acting in the land market, is that a public body can use its public law powers in order to strengthen its use of private law powers, to the detriment of those who have only private law powers.

There might be jurisprudence which has arisen about the interpretation of the relevant statute law or common law, and that jurisprudence might be important for determining the contents of citizen's rights. The principle of due process is very important in American practice and has been developed over the years: if a state body has taken executive actions without following 'due process', the citizen can appeal against that action. An equivalent in Dutch practice is called the principle of responsible government. This is a set of general principles (*algemene beginselen van behoorlijk bestuur*) to which *all* actions of a state body are supposed to conform, whether those actions are carried out under public law or private law. They are not incorporated fully into any codified law, so there is no official list of them, but they have full legal force (Needham 2014: 203–205).

Contradictions and Inconsistencies

With so many different ways of protecting citizens' rights, inconsistencies can arise. Two examples will be given as illustration.

In the Netherlands, appeals against the refusal of a permit used to be decided by a minister of the central government (*Kroonberoep*) not by an independent court. An injured party took her case to the European Court for Human Rights, which in 1985 decided in her favour (the *Benthem-arrest*). That was the occasion for the Dutch parliament to change its appeals procedure; now the final decision is taken by an independent court.

The second example comes from the UK. The European Convention on Human Rights was not directly enforceable in the UK until the UK government introduced its own version: the Human Rights Act 1998. Article 6(1) provides that "in the determination of their civil rights and obligations, everyone is entitled to a fair and public hearing by an independent and impartial tribunal established by law". That has important consequences for the procedures for spatial planning in that country. For example, appeals against a planning decision are not determined by an independent judiciary but by a 'planning inspectorate', which then advises a government minister, who decides on the appeal. The House of Lords (the highest court of appeal) has decided that this is not in breach of article 6(1), as the decision can be corrected by the right to an adequate and impartial judicial review (see above). Another consequence depends on the interpretation of civil rights in cases of spatial planning. For example, a third party who objects to the granting of a planning permission to another party and whose objection is turned down has, in the UK, no rights of appeal. Are that person's civil rights involved, and if so, how should they be protected? (Purdue 2004).

Implications for the Practice of Spatial Planning

The content of the laws protecting citizens' rights, together with the way in which those laws are implemented, have implications not only for the procedures which should be followed when making policy for spatial planning and when taking the measures to realise that policy. They can also have far-reaching consequences for the choice of the legal approach and for the content of the policy.

For example, if legal certainty and the associated condition of stability are given high priority, a planning authority will not want to undertake any spatial planning which depends on being able accurately to predict future events. And if the condition of impartiality is important (it is laid down in many national constitutions) then a planning authority will be careful about adapting planning regulations after consultations with land owners and property developers.

Main Conclusions

1. Political and ideological considerations lead to citizens being given rights which affect the way in which state bodies work, including the way in which spatial planning may be carried out.
2. Failure to observe such citizens' rights can lead to delays, court proce-dures, loss of support, unlawful decisions, etc.
3. The content of those rights and the relative importance given to them can affect both the procedures by which spatial planning is practised and the content of the planning policy.

Notes

1 That is the reason, for example, that under the current Dutch planning law there are very few formal rules about how a structure plan should be made; a structure plan does not bind anyone legally.
2 Note that predictability of the actions of a state body is not the same as predictability of the outcomes of the actions of that body. If a state body says in advance, "In the following circumstances, we will take no action", that is predictable, but not what the consequences will be.
3 Alexander (2002), too, uses it only for a limited type of case. He says: "A plan that does not enhance, or reduces, the welfare of the residents of the designated planning area is not in the public interest, unless the plan or its accompanying documentation demonstrates compelling public policy considerations in support of its provisions. This criterion is limited to cases where its terms are unambiguous, i.e. plans what are small and fairly simple, and where the residents of the planning area can be readily identified and are relatively homogeneous."
4 In the UK, a third party is every interested party except the applicant for the planning decision.
5 In the Netherlands, however, judicial review of statute laws is not possible.
6 The first principle is the 'doctrine of mixed law' (*gemengde rechtsleer*). This says that when a public authority acts under private law, it is subject also to the public law (van Wijk et al. 2005: 407). The second is the 'doctrine of the two ways' (*tweewegenleer*), where the two ways are public law and private law (van Wijk et al. 2005: 248). The current state of jurisprudence on this latter is that a public authority may use private powers to pursue its public responsibilities, only if the public powers

for this are inadequate, but not if the intention is to avoid the stringent procedures of the public law by using private law which has lighter procedures, nor if the private powers are used to contradict the established aims of that public policy (*doorkruisingsleer*).

References

Alexander, E. R., 2002. Planning rights: Toward normative criteria for evaluating plans. *International Planning Studies*, 7(3), 191–212

Alexander, E. R., 2007. Planning rights in theory and practice: The case of Israel. *International Planning Studies*, 12(1), 3–19

Booth, P., 1996, *Controlling development: Certainty and discretion in Europe, the USA and Hong Kong*. London: UCL Press

Campbell, H., Marshall, R., 2000. Moral obligations, planning, and the public interest: A commentary on current British practice. *Environment and Planning B*, 27(2), 297–312

Ellis, E. (ed.), 1999. *The principle of proportionality in the laws of Europe*. Oxford and Portland, OR: Hart Publishing

Ellis, H., 2000. Planning and public empowerment: Third party rights in development control. *Planning Theory and Practice*, 1(2), 203–217

Hartmann, T., Spit, T. J. M., 2018. Editorial: Dynamics of land policies – triggers and implications. *Land Use Policy*, forthcoming special issue.

Moroni, S., 2007. Planning, liberty and the rule of law. *Planning Theory*, 6(2), 146–163

Needham, B., 2014. *Dutch land use planning. The principles and the practice*. Farnham: Ashgate

Purdue, M., 2004. The human rights act 1998, planning law and proportionality. *Environmental Law Review*, 6, 161–173

Ramraj, V. V., 2004. Four models of due process. *International Journal of Constitutional Law*, 2(3), 492–524

Sorensen, A. D., Auser, M. L., 1989. Fatal remedies: The sources of ineffectiveness in planning. *Town Planning Review*, 60(1), 29–44

Wijk, H. D. van, Konijenbelt, W., Male, R. M. van, 2005. *Hoofdstukken van bestuursrecht*. Den Haag: Elsevier juridisch

5

Law and Policy Effectiveness and Efficiency in Spatial Planning

What This Chapter Is About

Effectiveness in realising the chosen planning aims, and the efficiency with which that is done, are important criteria when evaluating a legal approach to spatial planning. That evaluation requires making a causal connection between the measures taken and the predicted, or observed, effects. That connection is made by experience or by social science theories. The connection can be disturbed by intervening variables which change, and by unintended and unpredicted consequences. In practice, ends (the desired effects) and means (the measures taken) are mutually dependent. If achieving the effects is socially very important or urgent, different means will be chosen than if the desired effects are less pressing.

Introducing Effectiveness and Efficiency

Spatial planning is practised in order to achieve certain aims which the relevant government body has chosen. It may want housing to be provided in a certain volume at a certain location to reduce a housing shortage. Or it may want a new road connecting two cities, which would improve their accessibility. Moreover, it will want to do more than just *attempt* such things; it will want to be successful in doing so. Spatial planning needs to be effective, or as effective as possible. This holds for government policy and institutions in general: "Institutions are devices for achieving *purposes*, not just for achieving *agreement*. We want government to *do* things, not just *decide* things" (Putnam 1993: 8–9; emphasis in original).

Planning laws are made at the legislative level and give the public bodies at the administrative level the right to make legal land-use rules in order to structure and steer spatial development. Local authorities often use these legal rules as 'instruments' for spatial planning to achieve a desired effect, instead of, alongside or in conjunction with other instruments such as financial means (e.g. subsidies, incentives, funds) and communicative strategies (maps, visions, animations, images, verbal presentations and written statements). Since spatial planning should be effective, its instruments too should be effective. *Effectiveness* can be seen as an *absolute* concept with a dichotomous measurement scale: a policy is either effective or ineffective. Or it can be, and often is in practice, a *relative* concept: one policy is more effective than another or a policy becomes more or less effective over time.

This chapter is about the effectiveness of planning laws and legal rules. There is a lot of literature on the effectiveness of planning *policies*, at different levels of intervention (i.e. the strategic, tactical and operational level, e.g. Pressman & Wildavsky 1973; Mastop & Faludi 1997). However, *planning law and land-use rules* are rarely the objects of explicit evaluation, nor is there much literature that discusses such evaluations (for exceptions, see Faludi & Hamnett 1978; van Damme et al. 1997; Buitelaar & Sorel 2010; Buitelaar et al. 2011; Evers 2015). However, the effectiveness of law and rules can be assessed in a way similar to policies.

Policy makers, law makers and politicians are under budgetary constraints: they do not want effectiveness at all costs. Policy *efficiency*, which refers to the size of the input with which a given policy goal is achieved, is important as well. The policy with the highest output–input ratio is most efficient and to be preferred (assuming all other evaluation criteria are equal). Usually the input is measured in financial terms and relates to the costs associated with policy formation, implementation and monitoring. Some of these costs are referred to as 'transaction costs' (e.g. Webster 1998; Buitelaar 2004): there might be in addition costs made by the planning authority for acquiring and/or constructing land, buildings and infrastructure. Policy or legal efficiency is distinct from *economic* efficiency, which is discussed in Chapter 6. The former takes the policy goals as given, while in case of the latter the policy goal itself is also under scrutiny, as economic efficiency deals with the efficiency of resource (e.g. land) allocation, regardless of what the policy goal prescribes. Most of the issues discussed in this chapter apply to both legal effectiveness and legal efficiency, without always referring explicitly to one or the other.

Both effectiveness and efficiency can be predicted (to some extent) before a law or rule commences (*ex ante*) or measured after it has been terminated, withdrawn or otherwise stopped (*ex post*). In addition, laws and rules can also be evaluated while in action, during use and implementation (*ex durante*). An *ex durante* evaluation can help directly to improve legal practice and even the law itself (e.g. PBL 2010).

The Conformance and Performance of Legal Rules

When people think of effectiveness, they (implicitly) mean the level of *conformance* of the effects of policy with the aims of the policy. In the case of land-use rules, this means that it is desired that the spatial end-state be in conformance with the aims of the rules and the plans. If the land-use rules prescribe a certain number of houses of a certain type, the planning authority wants the end result to be in conformance with that.

The plan evaluation literature has introduced an alternative criterion: *performance* (Mastop & Faludi 1997). Here the object of the evaluation is not the spatial outcome or end-state, but the decisions by those who are addressed in, and affected by, the plan. In the case of performance, as developed in the planning literature, a plan is effective if it works through into, and influences the decisions of, other actors, regardless of whether their actions lead to conformity with the spatial end-state.

While performance may be a relevant criterion for strategic spatial plans, it is much less so in the case of laws and rules. In particular, performance is less relevant as an effectiveness criterion in planning systems that have a legally binding land-use plan or framework as their key planning document. In these cases, a planning application either conforms, and is approved (i.e. permission is granted), or does not conform, and is rejected. In systems where the central plan or framework is indicative, such as the English system with its 'local development framework', performance rather than conformance might be more appropriate, as development control takes these documents into consideration, as it does with 'material considerations' that are present at the point of that decision (Cullingworth & Nadin 2006). Growing insights may then lead to considering that the development framework is (partly) obsolete. Then, with a performance view on effectiveness, considering this framework and deliberately not conforming to it, is enough to constitute effectiveness, as it has influenced decision making.

Which Legal Rules?

In the previous chapters it has been said that the rules that are relevant for land use are made at different administrative or decisional levels. When rules are chosen and applied by a planning authority, that is the administrative or executive level. Those rules use legal instruments created at the legislative level. The question of the effectiveness and efficiency of rules can be asked about both levels. Are the legal instruments in themselves likely to effective and efficient? When a planning authority chooses and applies rules, is that likely to be effective and efficient? In this chapter, it is the latter question which receives most attention.

Also discussed in previous chapters is the question of 'rule-systems'. A separate rule is usually part of an aggregate rule-system, and separate rule-systems can be related to each other. Taking account of those relationships is part of the evaluation of a rule or rule-system.

Local governments may, and often do, pursue their goals with a *mix* of different rules and rule-systems. In Chapter 2, it was pointed out that there are 'packets' of property rights and land-use rules that apply to a particular plot of land, also referred to as the 'local regime of land laws' or the 'local user-rights regime', in which private- and public-law rules are combined. The packets or regimes determine the ownership and/or the content of property rights, which influence land use there.

For instance, if a local authority wants a run-down and contaminated site to be turned into a residential or mixed-use urban neighbourhood, it may decide to do the following. It buys the land under private law, like any other actor could do, possibly assisted by the 'threat' of compulsory purchase as a 'last resort' or *ultimum remedium*. It then remediates the contamination, demolishes existing structures, makes urban design plans, calculates financial feasibility, develops the land and prepares it for construction. Part of this process may be changing the content of the formal land-use plan: from allowing industrial use to allowing residential or mixed use. Additionally, it may decide to go one step further in detailing the desired end-state by adding conditions to the land sales, such as a particular share of affordable housing or a particular form of the public space, conditions which the party that wants to use the land for property development must satisfy. Such a process, where the planning authority uses a mixture of rules for one location, is common practice in the Netherlands. Together, this packet uses rules under private law (e.g. property law, contract law) and public law (e.g. spatial planning law, compulsory purchase law) to achieve a particular spatial end-state.

The Relationship Between the Application of a Rule and the Intended Effect

A planning authority which applies a particular rule or chooses a particular legal approach does so in order to achieve a desired effect (in any case, one would hope so; and see Chapter 4 for what the citizen must be able to expect from 'good governance'). The authority assumes that there is a cause-and-effect relationship between the measure and the desired effect.

The desired effect – a particular use of land and buildings – is brought about by people outside the planning authority: property developers, landowners, people choosing where to shop and live and work, people choosing where and when and how to travel, and so on. It is those actions which the measure should influence in a particular way. The planning authority wants, therefore, knowledge about the relevant cause-and-effect relationships.

That knowledge has two sources. One is experience with taking such measures in the past. When such measures have been taken in the past, what were the results? Experience must not be denied, but it is not always a good guide. What resulted in a thriving shopping centre at one time and place might result in empty shops at another time and place. A study of 'great planning disasters' (Hall 1982) is sobering and illuminating.

The other source is theories about how people create and use the physical environment (from human geography, from transport studies, from land economics, from environmental psychology, and so on). Those actions take place within a complex of circumstances, of which the planning measures are just one. If other parts of the context change, then so might the reactions to the measures too. That makes understanding the cause-and-effect change difficult.

Intervening Variables

'Intervening variables' are those that can 'intervene' between the cause (the planning measure) and the intended effects (the planning aims).

One of those is the ownership structure, which refers to the *fragmentation* of ownership and the *type* of owners that are present. If a planning authority wants an area to be redeveloped, and changes the rules to facilitate that, a fragmented ownership structure may hamper that and may therefore weaken policy/legal effectiveness or efficiency. This occurs likewise if the initial owners are not developers but farmers and individual households, who have no intention to leave or redevelop the area in which they live. These 'ownership constraints' (Adams et al. 2002) are intervening variables that have an impact on the end result and therefore on both the effectiveness and efficiency of land-use rules.

Another important intervening variable is the balance between demand and supply in local housing and real estate markets. This balance is time- and space-dependent. It is time-dependent because there are periods of market booms and periods of market busts, and various stages in between. It is space-dependent, as not every region has the human capital, economic structure and amenities to be economically successful: there are winners (nowadays the brain-hubs), areas that are in decline (e.g. the Rust Belt) and there are intermediary regions (Moretti 2012). The dynamics of regional housing and real estate markets reflect the regional economic situation.

In times and areas where demand is greater than supply, suppliers have greater freedom to produce housing, and demanders less freedom to choose between the supply. This allows, for instance, the supplier to pass on to the demander any price increases that result from land-use regulation (see Chapter 6). And providing affordable housing, and imposing a certain amount of affordable housing on property developers in land-use plans or building permits, is relatively easy (Allinson & Askew 1996). When supply – of housing, for instance – is greater than demand, demanders are in a strong position, and it is much more difficult to set goals and establish rules that are contrary to demand.

Predicting the Effectiveness of a Legal Approach

When a legal rule is applied, the intention is that it affects the behaviour of a private legal person in such a way that – directly or indirectly – a desired land use arises. It is important that the planning authority is clear about the expected relationship between the cause (the planning measure) and the effect (the desired land use), for that is a 'hypothesis' that it can use in order to increase the effectiveness and efficiency of its spatial planning. The hypothesis is used when choosing the legal approach (the *ex ante* evaluation) and, if the results of applying that approach are critically observed (the *ex post* evaluation), that might enable the hypothesis to be adapted for use in the following *ex ante* evaluation.

For example, the planning authority might make a land-use plan in order to realise a particular land use, specified by location and in more or less detail, and to be achieved by granting or withholding planning permissions. The effectiveness depends on the wish of many 'others' to develop and to use the land as planned. If that use is different from the existing use, the spatial planning will be effective if there is a demand to realise it: if not, it will be ineffective. In practice, this sort of spatial planning is often accompanied by discretion and flexibility (see Chapter 3). If demand is low, potential developers can be attracted by changing the content of the plan: if demand is high, willing developers can sometimes be persuaded to make extra contributions. In such cases, the plan can be effective, but not in realising the initial aims in all their details. Understanding the demand requires theories from human geography, etc.

Or the planning authority might want to realise a particular land use, specified by location and in more or less detail, by providing land and infrastructure. The actions which are taken by developers in order to realise that land use are encouraged by the planning authority providing some or all of the development land and infrastructure. Usually, this approach is accompanied by 'passive planning', whereby the development which is attracted will be allowed only if it conforms with a land-use plan. The effectiveness which can be expected is, however, greater than by passive planning alone, because it is made easier, and possibly cheaper, for potential developers to build. Because the planning authority has invested great sums 'up front', it will want to recover its expenses – quickly, too. So the planning authority will often use its discretion and flexibility; it will consider all sorts of deviations from the plan if that would induce others to develop. Once again, the plan might be effective, but not in realising the initial aims in all their details. The theories used for understanding how developers might react to the supply of land and infrastructure come from human geography, land economics, transport studies, etc.

Or the planning authority might decide to change prices, by levies or subsidies, so as to change the price incentives when people act voluntarily in their buying, selling and using land. The expectation is that actions subjected to a levy will be taken less, actions which are subsidised will be taken more. That is often done to influence transport behavior; for example, with subsidised public transport, parking charges, road pricing. The relevant theories for making such expectations are from economics (price elasticities of demand, etc.).

Difficulties When Predicting the Effects and When Measuring Whether They Have Been Achieved

Conflicting Goals

There are various challenges when evaluating effectiveness and efficiency. One of them is conflicting policy goals. Governments – societies, basically – want several things at the same time, sometimes things that are incompatible. On the one hand, they want to protect unique landscapes and biodiversity and, on the other hand, they want houses to be built to provide affordable shelter for citizens. Governments want low-density industrial estates to attract businesses and employment, while at the same time they want to contain urban development and limit urban sprawl. They want shops to be accessible by car and they want to limit carbon emissions and protect local air quality.

The difficulty of coordinating conflicting goals is enhanced by the fact that planning and policy making is usually a multi-level and multi-agency activity. Different government agencies at different government tiers have their own rationale, set their own goals and try to pursue them, partly through setting land-use rules. That can result in conflicting goals and conflicting rules.

Intervening Variables

Another challenge is the issue of 'intervening variables', discussed above, for these might not have been taken into account correctly or might have changed during the planning. As a result, conformance, or goal attainment, in itself is not enough to qualify land-use rules as effective – it is a necessary condition for rule effectiveness, but not sufficient. Other factors, or variables, than the rules themselves, might have caused or help produce conformance with the land-use rules. In that case, there is conformance with the rules, but it cannot be said whether the rules have been effective or not. Alternatively, unpredicted changes in those intervening variables might have prevented the predicted effects from being realised.

Unintended Consequences

Important in the light of effectiveness of land-use rules is the issue of 'unintended consequences', which are typical of complex social systems such as a city (Moroni 2015; Cozzolino 2017). In complex systems, there are many variables, often with non-linear relationships[1] between them, such as the relationship between the cause (for example, a particular land-use rule) and the effect (such as a spatial end-state) (Rauws 2017). Complex urban systems are, in addition, characterised by many feedback loops (both positive, reinforcing and negative, balancing feedback loops) (Forrester 1969). This limits the possibilities for explanation and prediction of specific events and actions, and opens the door to unintended consequences (Moroni 2015: 251–252). For instance, many cities in many countries have allowed out-of-town shopping centres that were convenient for their increasingly cardependent population. However, an unintended consequence was the decline of many (historic) inner cities.

Unintended and unpredicted consequences can have two effects that are relevant to the effectiveness (and efficiency) of land-use rules. First, unintended consequences may lead to a spatial end-state that is further away from the desired end-state. In other words, unintended consequences may lead to lower effectiveness. Second, a land-use rule may lead to unintended consequences that produce (negative) spatial side effects of that rule, instead or in addition to the effects it intends to produce. Negative side effects are particularly problematic if their harm is equal to or greater than the benefit gained from achieving the goal, the intended effects. It is difficult to measure such effects, but the possibility clearly calls for awareness of unintended consequences and negative side effects. For instance, land-use rules and zoning may constrain housing supply, and consequently inflate house prices, to such an extent that the benefits (e.g. saving green space) are deemed to be fewer than the costs (e.g. lower housing affordability). Such a 'regulatory tax' (Glaeser et al. 2005) may therefore be not only unintended but also undesirable.[2]

The Contextuality of Effectiveness and Efficiency

Lindblom states that in policy making, ends and means are not separately but simultaneously chosen (Lindblom 1959). If there is a great and urgent ambition to direct future urban development in a particular direction – for example, energy neutrality – then far-reaching means to achieve that are necessary. And, conversely, a government would set for itself aims that, given the means available, are not (totally) unfeasible. Thus, the effectiveness (and efficiency) of land-use rules depends on whether the available means (the legal rules and their limitations) are taken into account when determining the aims of the spatial planning.

This relates very much to the degree to which the planning authority can and wants to leave the future spatial end-state open. In Chapter 3, the distinction was made between the general aim of allowing a spontaneous order to arise and the particular aim of achieving a specific designed order, each with its own type of rule-system. Both can be effective, as they have different aims and different devices to achieve these aims. A detailed zoning plan is effective if the achieved spatial end-state is in conformance with the prescription of that end-state in the plan itself. An urban code is effective if it creates as many options for spontaneous action as possible (e.g. van Rijswick & Salet 2012), while also eliminating negative externalities such as noise nuisance, air pollution and hazards (Cozzolino 2017).

Effectiveness and Legal Certainty

The discussion so far has assumed that the aims – the desired effects – remain constant. However, in practice goals do not remain constant. Planning authorities might change their mind, either because of their developing insights or because of changing (external) circumstances. That may lead to a situation in which land-use rules are effective in achieving a goal that has ceased to exist. Then the planning authority might want to change its rules: the content of the planning policy, the legal approach taken, etc.

However, while that might improve the effectiveness as an internal goal, it might be ineffective in achieving an external goal (for that distinction, see Chapter 1), for the rules and rule-systems that governments in liberal democratic countries set themselves should provide legal certainty (see Chapters 3 and 4). This danger is great with detailed zoning plans. As detailed land-use plans become obsolete rather quickly, and have to be changed to accommodate changing circumstances, the (material) legal certainty they provide is primarily short-term; that is the time leading up to and just after granting the building permission. After that, legal certainty is low.

Main Conclusions

1. Spatial planning is practised in order to realise certain chosen aims, which are to be realised by bringing about a particular land use. It is therefore important, when choosing which planning laws to apply, and how, to predict beforehand whether that will be effective. And it is important to measure afterwards if that has been realised.
2. The connection between the measures chosen and the effects on the ground is often difficult to understand or foresee. Intervening variables can change, there can be unexpected and unpredicted consequences and the social system which results in land use is highly complex.
3. The knowledge and experience about those connections can be used not only to choose the most effective and efficient legal approach, but also to influence the choice of aims. If the likelihood of achieving a particular aim is predicted to be low, or to require a legal approach which might have undesirable side effects, it might be decided to change the aim.

Notes

1 A linear relationship is a proportional relationship between an independent and a dependent variable, which can be plotted on a graph with a straight line.
2 However, inflating house prices may be the deliberate aim. In this line, Fischel (2005) advances the 'homevoter hypothesis': local governments use zoning to constrain supply in order to increase the value of the assets of their constituents.

References

Adams, D., Disberry, A., Hutchison, N., Munjoma, T., 2002. Land policy and urban renaissance: The impact of ownership constraints in four British cities. *Planning Theory and Practice*, 3(2), 195–217

Allinson, J., Askew, J., 1996. Planning gain. In C. Greed (ed.), *Implementing town planning: The role of town planning in the development process* (pp. 62–72). London: Longman

Buitelaar, E., 2004. A transaction–cost analysis of the development process: A method for identifying transaction costs in different institutional arrangements. *Urban Studies*, 41(3), 2539–2553

Buitelaar, E., Galle, M., Sorel, N., 2011. Plan-led planning systems in development-led practices: An empirical analysis into the (lack of) institutionalisation of planning law. *Environment and Planning A*, 43, 928–941

Buitelaar, E., Sorel, N., 2010. Between the rule of law and the quest for control: Legal certainty in the Dutch planning system, *Land Use Policy*, 27(3), 983–989

Cozzolino, S., 2017. *The city as action. The dialectic between rules and spontaneity.* Milan: Politecnico di Milano

Cullingworth, B., Nadin, V., 2006. *Town and country planning in the UK.* London: Routledge

Damme, L. van, M. Galle, Pen-Soetermeer, M., Verdaas, K., 1997. Improving the performance of local land-use plans, *Environment and Planning B*, 24, 833–844

Evers, D., 2015. Formal institutional change and informal institutional persistence: The case of Dutch provinces implementing the Spatial Planning Act. *Environment and Planning C*, 33(2), 428–444

Faludi, A., Hamnett, S. L., 1978. *Bouwen en plannen in onzekerheid.* Alphen aan den Rijn: Samson

Fischel, W. A., 2005. *The homevoter hypothesis. How home values influence local government taxation, school finance, and land-use policies.* Cambridge, MA: Harvard University Press

Forrester, J. W., 1969. *Urban dynamics.* Cambridge, MA: MIT Press

Glaeser, E. L., Gyourko, J., Saks, R., 2005. Why is Manhattan so expensive? Regulation and the rise in house prices. *Journal of Law and Economics*, 48(2), 331–370

Hall, P., 1982. *Great planning disasters.* Oakland: University of California Press

Lindblom, C. E., 1959. The science of "muddling through". *Public Administration Review*, 19(2), 79–88

Mastop, H., Faludi, A., 1997. Evaluation of strategic plans: The performance principle. *Environment and Planning B*, 24, 815–832

Moretti, E., 2012. *The new geography of jobs.* Boston, MA: Houghton Mifflin Harcourt

Moroni, S., 2015. Complexity and the inherent limits of explanation and prediction: Urban codes for self-organising cities. *Planning Theory*, 14(3), 248–267

PBL, 2010. *Ex-durante evaluatie Wro: eerste resultaten.* Den Haag: PBL Netherlands Environmental Assessment Agency

Pressman, J. L., Wildavsky, A. B., 1973. *Implementation: How great expectations in Washington are dashed in Oakland.* Berkeley: University of California Press

Putnam, R. D., 1993. *Making democracy work: Civic traditions in modern Italy.* Princeton, NJ: Princeton University Press

Rauws, W., 2017. Embracing uncertainty without abandoning planning: Exploring an adaptive planning approach for guiding urban transformations. *DISP,* 53(1), 32–45. doi:10.1080/02513625.2017.1316539

Rijswick, M. van, Salet, W., 2012. Enabling the contextualization of legal rules in responsive strategies to climate change, *Ecology and Society* 17(2), 18. doi:10.5751/ES-04895-170218

Webster, C. J., 1998. Public choice, Pigouvian and Coasian planning theory. *Urban Studies,* 35, 53–75

6
Law and Economic Welfare in Spatial Planning

What This Chapter Is About

The starting point is the concept of economic welfare. Because spatial planning results in economic resources being used in particular ways, spatial planning can influence economic welfare. For that reason, it is necessary to be able to measure (changes in) economic welfare. Ways of doing that can be applied for the economic evaluation of planning projects and policies (cost–benefit analysis). When considering how to practise spatial planning in general so that it results in high economic welfare, it is necessary to be able to predict changes in economic welfare brought about by a particular legal approach. Most theories start from the idea that a market working under certain conditions will result in the highest possible welfare. If those conditions are not present (there are 'market imperfections'), the aim of public policy should be to create them by correcting for the imperfections. That correction can take two (legal) forms. One is by 'structuring' the market. This means introducing or changing private law rules, those rules which structure the voluntary interactions between people acting in the market, in such a way that the market works better. Another way of correcting for market imperfections is by 'regulating' the market. This means that a government body intervenes directly by forbidding some actions, by encouraging others, by itself taking actions.

Economic Welfare, Economic Resources and Spatial Planning

Economic theory uses the concept of economic welfare to refer to the value which people place on the goods and services produced, directly or indirectly, by a set of economic resources. The greater the economic welfare produced by a given set (e.g. all the economic resources in a nation-state), the better (all other things being equal, of course). A related concept is that of 'allocative efficiency': if the available economic resources can be reallocated (e.g. by spatial planning) in such a way that economic welfare is greater, the allocative efficiency, or the 'economic efficiency', is higher. If economic resources are being used in such a way that no other way could produce more economic welfare, this is referred to as the 'economic optimum'.

Correspondingly, economic resources are defined as those things which are used to produce goods and services which people want. People want goods, like cars and houses and food and clothes, and they want services, like mobility and music and a day in the country. Many of these things have to be produced, and for this resources are necessary, such as labour, expertise, raw materials, machines and land. There are also natural resources (i.e. not themselves produced) which produce valued goods and services directly and without any human processing. For example, a landscape produces a fine view, a river produces pure water. If they disappear, so does the value of what they offered.[1]

The land use in a particular location might contain natural resources, and it might contain land, buildings, infrastructure and so on which have been produced using resources (land, labour and capital) so as to provide goods and services which people want. In that respect, the land use can be regarded as an economic resource. Moreover, *using* land and buildings consumes economic resources, like travel time, and energy for travel and warmth.

The concept of economic resources is important because such resources are scarce. As a result, there is a limit to what can be produced with them. Scarcity is a relative concept: it means that people want more than is available. There is (at the moment) no scarcity of air to breathe, in some countries there is no scarcity of water to drink, in other countries no scarcity of building land. But often there is a scarcity of good housing, of playing fields, of road space, of attractive views. Scarcity is a relative concept in another sense too: if the things that people want change, then scarcity too can change. For example, when people used horses for transport, there could be a scarcity of stables; now that people use cars, there can be scarcity of parking spaces. If the things which people want have to be produced and people want more of them, then the limit to those things is set by the resources available for producing them. Are labour and machines going to be used to produce more houses or to produce more roads? Is land in a particular location going to be used to produce more playing fields or more housing? Are people going to use their time sitting in a car travelling to work, or relaxing at home? And if the things which people want are there already (like attractive views) but are threatened, are they going to be preserved if that would mean not having something else which people want: preserve a landscape which gives pleasure, or build offices and factories on it?

Spatial planning, which tries to influence how land and buildings are used, therefore affects how scarce resources are used. And in consequence, it affects economic welfare. Spatial planning can be, and often is, evaluated in that way. Which spatial plan or development project for a particular location would produce most economic welfare? If a change in the formal rules for spatial planning is being considered, would the new rules lead to economic resources being used in a better or worse way? Is there a general legal approach to spatial planning which would produce the highest economic welfare, the highest 'economic efficiency'?

Answering such questions requires that the concept of economic welfare can be made usable ('operational') in practice – can it be measured? It also requires that predictions can be made of the effects on economic welfare of taking a particular legal approach to spatial planning. These subjects are examined below, in that order.

Measuring Economic Welfare

In principle, it is possible to say whether economic welfare has *changed* as a result of a change in the way in which economic resources are used: you ask everybody who might have been affected by the change whether they feel better off or worse off. Of course, there might be some who feel better off and others who feel worse off. In that case you can do the following (again, in principle): you ask those who feel better off if they would be prepared to forgo some of their gain in order to compensate those who feel worse off; and you ask those who feel worse off if that transfer would compensate them for their loss. If the answer to both questions is 'Yes', then economic welfare has increased. Note: in such cases, economic welfare has increased even if the compensation has *not* been paid!

It is clear that this possibility would be impracticable in most cases. Nevertheless, it has received a lot of attention in welfare economics, and from this the concept of the 'economic optimum' has been derived. This optimum is reached when economic resources are being used in such a way that no alternative use would produce more economic welfare. That is called 'the Pareto optimum'. It was formulated in Pareto's *Manual of Political Economy*, first published in 1906, and has been refined greatly since (for an overview, see Little 2002, for example).

Those are interesting 'thought experiments' which, although not practicable in themselves, have led to some important ideas that *can* be applied to spatial planning. These are discussed later. First, more practicable ways of measuring economic welfare are discussed, ways which are easier than organising public meetings and enabling mass negotiations.

Market Outcomes as a Measure of Economic Welfare

The most common way of measuring economic welfare is by using *market outcomes* as a measure of the value of a good or service produced by the economic resources. The justification for doing this is as follows. People have money to spend, and they buy a range of products. The prices which they pay will reflect the relative values of all the various products which they buy, and the amounts that they buy will reflect how they want to spend their money. If fifty houses are built and sold for X, that has increased economic welfare by fifty X. More generally, the economic welfare in a society is the amount of a good or service produced, multiplied by its market price, summed up for all the goods and services produced. For example, the economic welfare of a nation-state is measured by the gross domestic product of that state. If resources were to be allocated differently, and if the sum of the market values of the new products was greater than that of the old products, economic welfare has increased.

The application of this to spatial planning can be illustrated as follows. Suppose that there is an open space of two hectares in an existing neighbourhood. Two different uses have been proposed for this: a neighbourhood park, or housing. How much economic welfare would each of the two uses produce? This is difficult to measure, but suppose all the people who might use it are asked how much they would be prepared to pay to have a park there. That would be a measure of how much they valued the 'services' from the park. How much economic welfare would the use of the park be for housing produce? The two values can then be compared.

Market Outcomes as Indicators of Economic Welfare

Under certain conditions, market outcomes might *not* be a good indicator of the effects for economic welfare. The most important conditions are the following, illustrated with applications to spatial planning.

External Effects

The external effects of an activity are physical effects and/or price effects, which are not experienced by the person who decides to undertake the activity. That person might therefore ignore them when making the decision. Of course, he/she might not: he/she might 'think of the others'. But there is no obligation to do so. Those effects can both harm others (negative external effects) and benefit others (positive external effects). If the effects are negative, the person causing them will not have to take account of them when prices are determined. A factory which causes pollution, for example, will not include in its production costs the costs of combating that pollution. So, the product prices might be too low. If the effects are positive, others benefit, and the production costs (and therefore the prices) are higher than if the producer had been able to profit from those benefits. In both cases, there are no market outcomes to measure the effects for the economic welfare of the external parties.

Monopolies

The mechanism which is supposed to make a market work efficiently is competition – between suppliers and between demanders. A monopoly excludes or reduces competition between suppliers. Land use can be affected by two sorts of monopolies, neither of which is absolute, but both of which can give an advantage to one supplier over other suppliers.

One sort is the monopoly granted by location, for the particular benefits granted by location cannot be easily reproduced. A piece of land next to an attractive park might be supplied for housing use. The supplier competes against owners of other plots of land next to the park. But one supplier might own all such plots. Moreover, owners of land further away from the park cannot supply the same locational advantages. This sort of locational monopoly affects, in particular, the supply of land near to transport facilities such as a railway station or highway interchange, and also land near to other locations desired for their views, natural amenity, fashionableness, availability of raw materials, etc. An example of locational monopoly that is particularly important for policy is when land is required to complete a road link, or to round off a new shopping centre. The owner of the land involved has a monopoly – no one else can supply the necessary land. The price the owner can demand for the land is higher than if there were competition. The price might be paid voluntarily; it reflects the value placed on it by user. But the volume consumed is less than if the price was lower. That is, the lack of competition affects the market outcome. The same situation can arise if a planning authority has a monopoly in the supply of development land.

The other sort of monopoly (called a 'natural' monopoly) arises when natural conditions are such that the *first* supplier achieves such a big advantage over possible competitors that he/she is not challenged. An example of this is a railway line, or a motorway, between A and B. The cost of that infrastructure is great, the first supplier takes all the demanders, and a possible competitor cannot start small but also has to build the whole infrastructure (the investment is 'indivisible'), while facing the risk of winning only part of the demand. Another example is a network for distributing gas and electricity. The first supplier can charge prices which include a monopoly profit.

Goods Where Users Cannot Be Excluded

For some goods, it is relatively cheap to enforce the rights of the owner to use them exclusively. An example is the right to exclusive enjoyment of one's back garden: if an unwanted and uninvited person enters the garden, the police can be called to reject the intruder. For other goods, enforcing the rights of the owner to exclusive use can be very difficult. Take as an example an attractive view. Someone might own the land which offers that view, but it is very difficult to prevent others from enjoying it too. If there is no management or rules to restrict use, it is an open-access resource (see also Chapter 2).

When consumption cannot be excluded, no price can be charged. So, even if the resources would give economic welfare, that cannot be measured by the market outcomes.

Goods Where Consumption Is Non-Rival

For some goods, there is rivalry in their consumption. This means that the more one person consumes, the less there is for others. Examples are food, housing, books, etc. For other goods, there is non-rivalry: even if one person consumes more of it, this does not reduce the amount others can consume. Examples of the latter are clean air, the view over a river, a big park. It is possible that a good for which the consumption was non-rivalrous becomes overused to the extent that consumption becomes rivalrous. An example is a public highway that suffers from serious traffic congestion – more use by some means less use by others.

When consumption is non-rival, charging a price for its consumption reduces the amount consumed and therefore the amount of economic welfare it gives. With lower (or no) prices, more would be consumed from the same economic resources and economic welfare would be higher.

Table 6.1 Private goods, public goods, club goods

	Users excludable	Users not excludable
Consumption rival	Private goods	Open-access resources which are overused
Consumption non-rival	Club goods	Public, or collective, goods

Source: derived from Webster and Lai (2003)

Private Good, Club Goods, Public Goods

Various combinations of excludable/rival are possible. Goods or resources where consumption is rivalrous and where consumers can easily be excluded are called 'private goods'; the opposite is 'public' or 'collective goods'. Examples of public or collective goods are most roads, most parks, police and national defence.[2]

There are goods (mixed goods) which fall between public and private, such as a small park. It is possible to put a fence around it and charge for entry. Consumption is non-rivalrous, but excludable – this is called a 'club good'. Sometimes consumers can only partially be excluded. Where exclusion is successful, the users can be required to pay, but where exclusion is difficult, people might be able to use the good without paying, such as those who jump onto busses or trains without a ticket. These are called 'free riders' – they enjoy economic welfare, but there are no market outcomes to measure this.

These combinations are summarised in Table 6.1 (derived from Webster & Lai 2003).[3] The 'Private goods' category is the only one in which market outcomes might give a good indication of economic welfare.

Imputing Market Prices

Where conditions are such that market outcomes are either absent or a bad indication of economic welfare, it is sometime possible to impute what the market price would be in the absence of those conditions (the 'non-market values').

The economic value of external effects, both positive and negative, can sometimes be estimated, so that all the effects on economic welfare can be included in one bookkeeping account.. Where people have free access to public goods and to resources in the public domain, the value which people attribute to that can be estimated by questionnaires; they measure 'the willingness to pay'.[4] Sometimes, what are called 'shadow prices' can be estimated. For example, the shadow price of a nature reserve which is lost might be the cost of replacing the natural values in that reserve in another location. And sometimes econometric analysis reveals how people value certain non-traded goods and services.[5]

The Effects of the Distribution of Welfare on Market Prices

The relative prices that arise in markets reflect the distribution of incomes and wealth among buyers. 'Individual preferences, as revealed in market behaviour, are a function of [people's] income and wealth' (Ogus 1994: 58). Take the example discussed above of the open space of two hectares in an existing neighbourhood, which could be used for a park or for housing. How can we measure the welfare which the space would give as a park? If the households living around it are rich, they will be prepared to pay a lot to get the space allocated as a park. If the households are poor, they will want to pay less. The economic welfare given by the park is higher in the first case than in the second. But common sense would suggest the opposite: poorer people have smaller gardens and less opportunity than richer people for enjoying open space elsewhere, so the park should give poorer people more welfare. Take another example. Suppose that a route is being sought for a new road. One possible route would require demolishing ten big and expensive single-family houses; the other possible route, the demolition of forty small and cheaper single-family houses. If the value of the ten big single-family houses is greater than of the forty small single-family houses, then demolishing the forty small single-family houses would give greater economic welfare than demolishing the ten big single-family houses.

Using market prices to measure economic welfare implies that the initial distribution of wealth and income is accepted.[6]

Cost–Benefit Analysis in Spatial Planning

Suppose that there is more than one way of applying laws (more than one legal approach) when tackling a particular planning issue. One of the considerations is this: which approach would lead to higher economic welfare? An answer can be sought by making a cost–benefit analysis of the results expected from the different approaches. Which approach would produce the higher welfare?[7] The welfare effects are predicted and given a value (volume * price). If the analysis includes not only the direct but also the indirect effects, it is sometimes called a social cost–benefit analysis. Nowadays, such analyses are routinely applied.

Special attention needs to be paid to the position of land values in cost–benefit analysis. If it is claimed that the aim, or one of the aims, of spatial planning is to increase economic welfare, then this can be interpreted as: increase the economic efficiency with which land and buildings are used. And this, in turn, can be interpreted as: that land use which gives the highest property values is the best. However, in practice, property values are not included in cost–benefit analysis (or not directly – see below). There are two reasons for this.

One is that a change in property prices caused by realising a spatial plan is often associated with changes in land prices elsewhere. For example, if the shops in a town centre are improved, this might cause land prices there to increase, but if that improvement leads to fewer people shopping in competing centres, land prices in those competing centres will fall. That is to say, there are external effects outside the plan area – there is substitution and no net gains.

The other reason is more fundamental. Suppose that two alternative locations for a new development are being compared. They will have different effects on traffic movements. The cost–benefit analysis will try to quantify those effects – how many movements, over what distances, etc. – and will try to put a monetary value on those effects. Those differences in traffic movements might have effects on property values. For example, the location which is best located for work and recreation has higher house prices. However, those higher prices are the result of shorter journeys. Including *both* the money values of the differences in journeys *and* the difference in property values would be double-counting. If property values are included in a cost–benefit analysis, it is only indirectly, when they are used to impute values to something which is not directly traded in the market, such as enjoyment of landscapes (see above).

Cost–benefit analysis as a way of comparing the effects on economic welfare of alternative approaches is subject to certain limitations; these are the attribution of prices to some of the effects might be debatable (see above), and the existing distribution of wealth and income is not questioned.

Evaluating the Economic Welfare Effects of a Legal Approach to Spatial Planning in General

The Need for a Theory

Cost–benefit analysis can be used to estimate the economic welfare effects of alternative ways of tackling a *particular* planning issue. There is another, more general, question which can be asked when evaluating the welfare effects of legal approaches to spatial planning. What are the economic effects of a particular way of practising spatial planning *in general*? This question is important because it is often suggested that the customary way of spatial planning (with land-use plans, planning permissions, etc.) depresses economic welfare, and that if spatial planning were practised in a way which does not regulate as much, welfare would be higher. Answering that question requires a *theory* for predicting the effects of policy decisions on economic welfare. So what is the underlying theory?

The Perfect Market and the Economic Optimum

A commonly used theory starts from the idea of a perfect market (Dutta 1994). The perfect market will realise, on its own and without government intervention (such as by spatial planning), maximum economic welfare. This market would work without rules from the state; it would be a 'free market'. This has been called the First Fundamental Theorem of welfare economics (Dutta 1994: 16).[8]

A weakness of this 'first fundamental theory' is there never can be a market free of state rules. Markets cannot work without rules of some sort or another. This has been shown in Chapter 2 with respect to markets in land and buildings: there need to be (private law) rules to create and maintain (economic) value and to facilitate exchanges between actors in the market. 'Markets need the state', say Webster and Lai (2003: 52); and, less prosaically, 'If the market is the dance, then the state provides the orchestra and the dance floor' (Lindblom 2001: 102). For this reason, there can never be a market without rules. Moreover, there is a variety of possible (private law) rules for enabling markets to work. As a result, 'the one and only free market' does not exist – there is a wide variety of possible markets, each working under different rules. And it can be expected that the welfare produced by the use of economic resources exchanged under market conditions will depend on the content of those rules. If it is argued that the market should be left to operate 'on its own', the argument should include attention for the private law rules which should be in force in order to enable this. (See 'Market Failures Corrected by Structuring Markets' below.)

The Market as the Starting Point

Nevertheless, it is not an implausible idea that, if economic resources and their products are exchanged in the market, they will be used efficiently. A market works by people making choices about how to employ their resources and what to buy and sell, so those choices might result in what people want. Moreover, it is not a bad starting point, for it is inconceivable that there could be production of goods and services (including the production of the land use) with *no* contribution from markets (Lindblom 2001). It is almost impossible to imagine a situation in which *all* decisions about land use are taken by a state body, to the total exclusion of free exchange between citizens and private organisations. The market cannot be excluded from decisions about real estate, even if it were considered desirable to do so.[9]

The idea of the market as starting point has a long history, being first taken by Adam Smith in his *Wealth of Nations* (1776).

> Give me that which I want, and you shall have this which you want
> . . . and it is in this manner that we obtain from one another the far
> greater part of those good offices which we stand in need of. It is not
> from the benevolence of the butcher, the brewer, or the baker that we
> expect our dinner, but from their regard to their own interest.
>
> (Book I, Chapter II: 6 et seq., 1866 ed.)

The market provides an 'invisible hand' which coordinates individual actions so that together they produce a certain use (which might or might not be the *best* use; Smith said nothing about that) of scarce resources.

Much applied economics is based on the idea of the market producing economic welfare, and from this notion public policies are derived. There is an academic discipline called 'law and economics' which uses economics to analyse the effects of *private* law (see, for example, Cooter & Ulen 2004).[10] Predicting the effects of *public* law is a well-developed study – micro-economic policy and macro-economic policy. Both are included under the academic discipline of welfare economics. How can that be used to predict the effects of law and public policy on economic welfare, as defined above?

Market Imperfections

With the idea of the market as a starting point, what can be said about policy (including spatial planning) which aims to increase economic welfare? The theory of the perfect market producing maximum economic welfare draws attention to what are called 'market imperfections' or 'market failures'. These are conditions which can *lower* the economic welfare resulting from market transactions. It follows, according to this theory, that if those imperfections are corrected, economic welfare will increase. Even if the theory of the perfect market cannot be tested, the arguments about market imperfections are highly plausible.

Some of those imperfections are the conditions investigated above, conditions which result in *market outcomes* not being a good measure of economic welfare; namely, external effects, monopolies, non-excludability, non-rivalry. An additional imperfection is that caused by high transaction costs. We now investigate the effect of these market imperfections on economic welfare itself, with illustrations from land-use decisions.

External Effects

External effects are sometimes called 'spillovers' and, when they are negative, 'social costs', but that latter term is misleading. It suggests that there is a relationship between the causer of the external cost and 'the society'. In most cases, however, it is relatively easy to identify who suffers from the external costs. External effects are usually a relationship between individuals and individuals (Bromley 1991: 19).[11]

It will be clear that if someone makes a decision which takes no account of the effects on others, that choice will not necessarily improve economic welfare.

Land-use decisions often cause externalities. Well-known examples are noise nuisance, smells and fumes, shops attracting each other, expensive housing attracting other expensive housing, industry disturbing recreation. The geographical dimension is important: the externalities are caused by an activity in location X and the effects diminish as you get further away from X. For this reason, the physical effects caused by an activity can be an external effect under some circumstances, and an internal effect under other circumstances. If the activity which causes the effect is placed in the centre of a very big plot of land, the externalities will be negligible on adjacent plots: all effects, both positive and negative, are internal; that is, within the extent of the land owned by the decision maker.

Monopolies

If someone exercises his/her monopoly power, this can reduce economic welfare, for the monopolist can reduce supply and charge higher prices than if there was full competition. A housing developer, for example, who has acquired most of the development land in a town, can build those houses at a rate which suits him/her best, and restricting that supply can push up prices.

Non-Excludability

If people cannot be excluded from using a resource, no price can be charged, and people might use it extravagantly (up to the point when it gave them no more welfare/utility). Moreover, that might lead to the depletion, even exhaustion, of the resource; then it can give no more welfare. Economic welfare, in the longer run, would be higher if use were rationed. And if, because no price can be charged, the good is not produced even though consumption would give economic welfare, then that welfare is not realised.[12]

Non-Rivalry

If consumption is non-rival, and the producer of that good charges a price for its use, consumption will be reduced, even though more (free) consumption would not reduce the welfare of the paying consumers. Consider, for example, a small park. This could be provided privately, whereby users had to pay. Then, fewer people would use it than if it were provided free. Yet, with free access, more people could use the park without reducing the existing users' enjoyment.

High Transaction Costs

If you buy a good for the price X, the cost to yourself is usually more than X. And the producer who sells that good for X has to incur considerably more costs than the direct production costs. The difference in both cases lies in the costs of making the transaction. These are usually divided into: search (or information) costs; bargaining costs; enforcement costs (Cooter & Ulen 2004: 92).

With respect to the activity of physical development, many transaction costs can be identified and fitted into the above classification. If someone considers buying a building, he/she makes forecasts of what might happen insofar as that might affect the value and costs of that building in the future. These are search costs. If the building is bought, there might be a lot of bargaining. And if it has been bought under certain conditions, someone will have the costs of checking that those conditions are being met. These are enforcement costs.

Coase (1960) argued that many transactions do not take place, although the outcome would have been advantageous to all the parties (thus increasing economic welfare), because the costs to the parties of making the transaction would have been greater than their gain. This can be illustrated with an example from urban development. There is a block with mixed uses in a city centre. The capital value of that is M. If that block were redeveloped, the value (net of the development costs) would be M + N. The gain (N) could be divided so that all the parties involved agreed to the scheme. Why does it not take place? Because the transaction costs are greater than N. These are the costs of researching whether the scheme might be profitable, the costs of discovering who has what rights, the costs of negotiating with all the parties, the costs of finding a financier and convincing him/her, and so on. In land-use planning, this is sometimes referred to as a missed opportunity – it is missed because coordination is so difficult that market parties on their own cannot realise it (see, for example, Harrison 1977: 71–74).

The Theory of the Second Best

According to those implications from the theory of the perfect market, if those market imperfections are corrected, economic welfare will increase. Two ways of correcting for the imperfections directly or indirectly will be investigated below: 'structuring markets' and 'regulating markets'. First, however, it is necessary to note that the practical conclusions (that is, correct for market imperfections) are not undisputed. The doubt is expressed in what is called 'the theory of the second best' (Lipsey & Lancaster 1956).[13] According to this, if there is more than one market failure, correcting for one of them (by, for example, spatial planning) will not necessarily result in a higher economic welfare.

Two Legal Approaches to Correcting for Market Imperfections

That theory about free markets, market imperfections and economic welfare is now applied to two sorts of legal approaches in general, varieties of which were distinguished in Chapter 1. One of the two approaches involves a planning authority using rules to impose restrictions on how people act in the land and property markets. Here this is called the approach of regulating markets. The second approach is called structuring markets, and leaves market actors as free as possible, working under private law rules which are intended to correct for market imperfections.

Note that this comparison of the two sorts of approach tackles the issue of which approach would result in land uses that would give the highest economic welfare. The issue is not which approach would best give the land use which a planning authority wants to realise in order to achieve certain aims. The *external* aim of economic welfare is central: if there are any *internal* aims (e.g. a more attractive town centre, more housing), they are irrelevant to this comparison. The comparison starts with the approach of structuring markets.

Market Failures Corrected by *Structuring* Markets

The approach of 'structuring markets' starts from the idea that legal persons acting in their own interests will produce the highest economic welfare, if the conditions are correct. And the conditions should be such that people can act in their own interests as freely as possible; that is, without direct public intervention. That can be done if government creates private law rules (see Chapter 2) which provide a structure for voluntary actions in such a way that market failures do not arise. Neither planning laws, nor any other public laws, would be applied. Because it is (usually) the national government which makes private law rules, planning authorities would then do no more than 'let the market work'.

Property rights are one form of private law rules, and many such rights have been created precisely to require the producer of negative external effects (nuisance) to take account of those, either by requiring the producer to pay compensation (liability) or by giving the victim the possibility of requiring the producer to stop (an injunction). Such property rights require 'the market' to take account of external effects. Structuring markets involves introducing property rights which do that well, possibly in conjunction with contract law.

External Effects and Transaction Costs

Take the example of negative external effects. These arise, it is argued, because those harmed by the actions of others have no right to object. There are 'incomplete property rights' or 'property rights ambiguities' (Webster & Lai 2003: 95). Those living near to a polluting factory do not have a right to unpolluted air, so they cannot go to court to have that right enforced. And even if they could, the transaction costs of doing so – of organising all those harmed by the pollution so as to take coordinated action – would be so great that it might deter them from taking action. Alternatively, those harmed by the pollution could bargain with the polluting factory to get compensation. But how do you get all those harmed to act together? This, too, is a question of property rights and transaction costs.

If those conditions ('complete' property rights and zero transaction costs) were present (created by the legislator), those suffering *negative* external effects could require the 'damagers' to pay compensation, and negotiations over that would be costless. If someone was considering an action which would cause *positive* external effects, he/she could negotiate with those who would benefit, so that those benefiting would agree to pay some of the costs of the action – and, again, those negotiations would cost nothing. A 'market in externalities' has been created, and 'the problem of social costs' would disappear (Coase 1960; also see Pearce, 1981, who was one of the first to compare tackling external effects either with development control through public law or with extended private property rights).

With lower transaction costs, there would be fewer 'missed opportunities'; 'barriers to trade' would be reduced. Lower transaction costs could also help if desirable actions are being held up because of imperfect information. In such cases, the market could be 'assisted' if a public body set up a cadastral or public land registry, or carried out and published surveys about population change or housing wishes, or made projections about how the economy will change. This is 'providing land markets with subsidized strategic intelligence' (Webster & Lai 2003: 26; also see Buitelaar 2004). As Cooter and Ulen say (2004: 97), lowering transaction costs 'lubricates' bargaining.

Non-Excludability

If potential users cannot be excluded under the existing rules, perhaps those rules can be changed. It is sometimes suggested, for example, that open-access resources be privatised: many people have concluded this from Hardin's analysis of the tragedy of the commons (1968). And it has in some cases been realised: some commons have been enclosed, public shopping streets have been replaced by private shopping malls. (But such a solution is technically difficult, if not impossible, for some open-access resources, such as fishing on the high seas.) With a change of rules, open-access resources can be – and have been – made into club goods. For example, a small local park which had been provided by the municipality and was open to all becomes a park to be used only by the local residents, who also maintain it (Webster 2007). The Homeowners Associations under which so much private housing in the United States is developed are another example: the streets and the open spaces are maintained by the residents and used by them only (Nelson 2004).

Recently, technological innovations have made it possible to regulate the use of some public goods: people can be required to pay for access. The streets of central London, for example, could be used freely by any driver until the introduction of a 'congestion charge' in 2003. The charge was introduced in central London because of traffic congestion, which meant that 'consumption' of the public highway had become rivalrous. (For a fuller discussion, see Webster & Lai 2003: chap. 6.)

The general conclusion is that, with different property rights and lower transaction costs, many market imperfections would be removed; that is, markets could be 'restructured' by changes in private law so that the market itself would achieve higher economic welfare. The classic treatment of this is by Coase in his paper 'The Problem of Social Cost' (1960), in which he argued that if there were no transaction costs and if the rights of all parties were well defined, the market would produce maximum economic welfare on its own; that is, without regulation from outside.[14] The 'normative Coase theorem' says: 'Structure the law so as to remove impediments to private agreements' (Cooter & Ulen 2004: 97).[15]

Correcting for Market Failures Corrected by *Regulating* Markets

The approach of 'regulating markets' is different. Public law rules should be created, whereby a government body can 'intervene' in private actions, prohibiting some, encouraging others, so that the result is the land use which would have resulted if there were no market imperfections. The laws for doing this are discussed in Chapter 3. Most spatial planning is practised in this way. It has been for hundreds of years, and it still is.

The starting point is that 'the market' should be the 'default' way in which goods and services are supplied, demanded and distributed, but where there are market failures, the government should intervene to correct for these. The classical treatment is by Pigou, in *The Economics of Welfare* (1932). His argument was that, if the conditions were not those required for optimal allocative efficiency – if there were 'market failures' – then it was the task of the state to correct for those market failures in such a way that the allocative efficiency would be better than the market itself would achieve under the 'imperfect conditions'. The corrections could have the form of taxes, or subsidies, or restraints, or state production, or state coordination.

Pigou himself applied this to land-use planning:

> It is as idle to expect a well-planned town to result from the independent activities of isolated speculators as it would be to expect a satisfactory picture to result if each separate square inch were painted by an independent artist. No "invisible hand" can be relied on to produce a good arrangement of the whole from a combination of separate treatments of the parts. It is, therefore, necessary that an authority of wider reach should intervene and should tackle the collective problems of beauty, of air and of light, as those other collective problems of gas and water have been tackled.
>
> (Pigou 1932: 195)

Since then, many others have worked out further the practical implications of Pigou's ideas (e.g. Ogus 1994: chap. 3), including their implications for spatial planning (e.g. Harrison 1977; Heikkila 2000). Indeed, this is one of the justifications most often given for spatial planning: it is 'intervention' in the market, to correct for market failures.

Correcting for External Effects

If there are external effects, such as a factory chimney which emits polluting smoke, the Pigovian correction is that the state should tax the factory, or prohibit the emission of dirty smoke. Much spatial planning has the form of preventing or reducing possible negative external effects by regulating the locations of the relevant land uses (see Chapter 3). One of the oldest instruments for spatial planning is the zoning (or land-use) plan, the aim of which is to reduce external effects by separating land uses which might cause them from land uses which might suffer from them.

There are other possible regulatory measures. For example, if the external effect is *positive*, boundaries can be deliberately drawn so as to include them: this is called 'scoping'. This allows the person causing the effect to reap all of the benefits, because they fall within his land. Take the example of a multi-storey car park in the town centre. Building this will benefit not only the owner, but also nearby retailers: if the car park is built as part of a new shopping centre, the one developer can recoup that externality. Webster and Lai (2003: 155) give, as other examples of scoping, shopping malls, business parks and university campuses. 'Proprietary communities' (Pennington 2002: 91) have been created. If the external effect is *negative*, scoping can make it internal. For example, if a large factory is required to locate the most noisome activities in the centre of its (big) plot of land, the only person suffering from it is the one who causes it. Or suppose an office block is built, which results in many more people wanting a parking space. If they may freely seek that space in the surrounding streets, that physical effect is external to the decision making about the office development. If it is not permitted to build an office in that location without making provision for the demand for extra parking, the physical effect is internal to the decision making. The externality has been internalised.

Spatial planning can stimulate or require scoping by coordinating the development decisions at a spatial scale higher than the separate building plot. This is the planning practice of 'coordinated urban development'. For instance, in the Dutch case of Almere Oosterwold, landowners are prohibited from causing negative external effects on their plots. Consequently, their plots need to be large enough to internalise them.

Correcting for Monopolies

Some plots of land have a location which is both unique (or very scarce) and desired, such as the land required to complete a road link, or a plot crucially located within a big development project (a locational monopoly). The owner then has a monopoly over the services which that plot can render. The Pigovian correction is that the state should purchase the plot, if necessary compulsorily. In practice, roads, railways, canals could not be constructed without the state using (the threat of) expropriation.

Where there is a natural monopoly, the government often regulates how it is to be used. For example, in order to encourage investment in expensive infrastructure such as a rail line, it can grant a monopoly to one supplier (a franchise), under conditions to prevent that supplier abusing its monopoly powers.

Correcting for Non-Excludability and Non-Rivalry

The consequences of non-excludability are that no private suppliers are interested in providing the resource. The consequences of non-rival consumption are that private suppliers do not know how much of a good or service to produce. The Pigovian correction is that a state body should provide the resource (a road, a park, etc.) as a public good. If non-excludability should lead to over-use, a government body should regulate consumption by requiring the user to apply for a licence. Examples are over-fishing regulated by fishing quotas, or the extraction of groundwater and minerals regulated by permits.

Imperfect Information and Coordination Difficulties

Suppose that a developer is considering building houses in a particular town. He/she does not know for certain how many of those houses he/she will be able to sell, nor at what price. Another developer, faced with the same uncertainty, might make a different decision. It is unlikely that both decisions would give the best allocation of resources.

Take another example. There is a street of run-down houses. One person (Mrs X) considers improving her house, which would cost M. If others do not improve their houses, the value of Mrs X's house will increase by less than M, for the other houses in the street remain unattractive. If all others do improve their house, at the cost of M, the value of each separate house will increase by more than M, for the whole street has become more attractive. But how can Mrs X get all those other people to do what is in their individual interest, but which works out only if everyone agrees to take part?

In terms of the discussion above, there might be high transaction costs, which prevent a good solution being realised. The Pigovian correction for this (although Pigou himself did not use the term) is that the state should provide overviews and forecasts, and so provide a more certain framework for investment decisions. And if a state body (such as a local government) tries to coordinate the decision making over a development project, this can reduce bargaining costs. The Pigovian correction for this is that the state takes on the role of neutral and trustworthy coordinator. However, regulation always brings its own transaction costs with it (i.e. costs of enforcement).

Regulatory Failures

Regulation is supposed to correct for market failures. But it is not only the market which can fail, as regulations can also do so. There is a wide literature and much empirical investigation on what has been variously called 'public failure' or 'government failure' (e.g. Dolfsma 2013), or 'non-market failure' (Wolf 1979), or 'regulatory failure' (Ogus 1994: chap. 4). Not all of it is relevant to this chapter. The concept of regulatory failure discussed here refers to the failure of public interventions to regulate the market so that it achieves higher economic welfare than without the regulation. The concept of government failure is wider: it includes other sorts of failure also, such as the failure of public policy to achieve its stated goals (the question of effectiveness; see Chapter 5). Relevant to this chapter are two questions. What is the success of public policy in correcting for market failures in land use, such as external effects, monopolies, non-excludability, non-rivalry, high transaction costs? What unintended and unwanted side effects can this bring about?

The literature focuses on the following points:

- Does a government body have sufficient and correct information to do that? Market actors are often better informed.
- What are the costs of regulation; that is, the costs of the public administration, and the costs to those regulated (bureaucracy, delays, etc.)?
- Are politicians and public officials correctly motivated? Do they let themselves be captured by private lobbies and interest groups? (This is the subject of what is called 'public choice theory', Tullock 2008.)
- Does public policy (in this case, spatial planning) create the conditions for rent-seeking? This arises when the public policy results in, or strengthens, monopoly or oligopoly, whereby there are differences in prices which cannot be reduced by competition and which, therefore, benefit some and not others (Ogus 1994: 72 et seq.).
- Does the public policy result in higher prices (e.g. by restricting supply) which are not the aim of the policy?

Just to ask these questions is a warning against assuming that spatial planning by regulating markets will, by trying to correct for market failures, produce higher economic welfare than the market itself would.

Choosing a Legal Approach to Spatial Planning

The statement that spatial planning can contribute to economic welfare is indisputable; also that economic welfare is important for the society. Much has been written about the relationship between economic welfare and spatial planning and, out of this, two ideal types of legal approach have been identified: creating private law rules which *structure* markets, and creating and applying public law rules which *regulate* markets. For both, claims have been made about the consequences for economic welfare. How relevant are those arguments for practice? That depends on the answers to two questions:

- How well can economic welfare be measured?
- How reliable are the theories which predict the effects of the two legal approaches to spatial planning?

Answers are sought below.

How Well Can Economic Welfare Be Measured?

Observable market outcomes can often be used to give an indication of economic welfare. However, under some conditions, market outcomes give a poor indication, and sometimes they are not observable. It is sometimes possible to correct for this (see the discussion above). But the logical limitations to this must be recognised.

How Reliable Are the Theories Which Predict the Effects of Spatial Planning?

Two legal approaches to spatial planning have been identified above: structuring the market and regulating the market. And much has been written by those who argue that one approach is better than the other (e.g. Grant 1988; Pennington 2002). However, it is not possible to conclude using economic theory which of the two approaches would result in the highest economic welfare. As Pejovich (1997: 148) says, 'The choice is between two or more imperfect systems.'

The weakness of many of the arguments presented above is that they rest on the assumption that there is something called 'the economic optimum' – the greatest economic welfare that can be produced with the given set of economic resources – whereby public policy (here, spatial planning) should be directed to achieving that. This attempt to find a policy which would achieve the optimum has been called 'the Nirvana fallacy' (Furubotn & Richter 1991: 12), which can be described using another mythical term as 'the search for the Holy Grail'.

An additional weakness of the economic analysis is that it has been carried out from the starting point of markets; namely, that it is markets which coordinate the exchanges between anonymous buyers and sellers, exchanges which result in land and buildings being distributed, changed and used. This, however, is incomplete, for many economic resources are voluntarily exchanged in other ways, too: by mutual trust and self-organisation. In those other ways, decisions are made about how to use economic resources, how much and at what price. This, also, has effects for economic welfare. But the theories from welfare economics used to predict those effects, and discussed above, are seldom applicable to such exchanges.

How Comprehensive Can an Economic Evaluation Be?

There is one final point. The legal approach taken to spatial planning affects economic welfare, because it influences how economic resources are used. In addition, it might have lots of other effects, such as on justice (Chapter 7) and legitimacy (Chapter 8). Then another question arises. What can and what should be included in the concept of 'economic welfare'? Can *all* the effects of spatial planning be included? It would be very convenient if they could, for that would mean that all spatial planning projects, and all legal approaches, could be subjected to one and the same evaluation; namely, an economic one. It is for this reason that the question has been much studied.

Moral Issues

The complication arises because moral issues are sometimes involved. There used to be a flourishing market in slavery, and the effects on economic welfare could easily be measured (the market outcomes). But slavery is now forbidden. A landlord might be able to get higher rents for his/her apartments if he/she refuses to let them to ethnic minorities, and the welfare effects of letting with that discrimination can be compared with the effects of letting without that discrimination. Such discrimination is now forbidden in many countries. Or suppose that it were profitable to build a shopping centre on a military war cemetery. Would that be seriously considered? A similar question arises if using economic resources would affect something which has an 'intrinsic' value – it has a value in itself, apart from any value which might arise if it were traded. Many people consider some aspects of nature to have an intrinsic value, such as the survival of particular species. In such cases, it can be decided that the resources should not be used in that way, so there are no market outcomes. And even if it were possible to trade them in the market (e.g. the illegal trade in protected species), it might be decided that the market outcomes are irrelevant for decisions about economic welfare.[16]

Discounting the Future

Suppose that a development project would use certain resources (such as fresh water) in such a way that future generations might be seriously hindered. It has been suggested that the interests of future generations can be taken into account economically by discounting them at a low interest rate. Using an interest rate of 2 per cent as an example, the effect of this is that anything which happens after fifty-seven years is only one third as important as if it happened now.[17] About this, Bromley (2001) says: "Relying on the realm of calculation to diminish (to discount, both literally and figuratively) the standing of future persons violates behavioural norms located in the realm of sentiment."

Political Issues

The objections to trying to include *all* the effects of a policy measure in one grand economic evaluation are also political; namely, that it takes discussions out of the hands of politicians and put them into the hands of 'the experts':

> by developing a sophisticated concept of total economic value, economists can embrace such concepts as existence value, altruism and stewardship, and assess preference in relation to them. It is probably with that final step that non-economists feel most uneasy, even if it means opting instead for the safety of paternalism and the loose fall back of democratic accountability.
>
> (Grant 1988)

> Maximum national income . . . is not the only goal of our nation as judged by policies adopted by our government – and government's goals as revealed by actual practice are more authoritative than those announced by professors of law or economics.
>
> (Williamson 1999: 318, quoting Stigler 1992: 459)

And: "Don't think economics, think society", said Lindblom (2001: 19). In particular, when a legal approach to spatial planning is being chosen, it is not just the effects for economic welfare which should be taken into account.

Distributional Issues

Finally, it must not be forgotten that the distributional effects of spatial planning can be very important (also see Chapter 7). Sometimes spatial planning aims to change the distribution deliberately, such as when taxpayers' money is used to upgrade a housing area with poor facilities – welfare is being distributed from the general taxpayer (who would otherwise use that money for something of his/her own choosing) to the households in the improved area. An unintentional redistribution is when a new road is built in an attractive landscape: the road users benefit, at the cost of those who previously enjoyed the landscape. However, the way in which economic welfare is distributed among the citizens of the society is not included in the concept of economic welfare.[18] Nor is the fact that market prices are affected by the distribution of income and wealth. Suppose that there is a change which increases the welfare of some, and reduces the welfare of others. As long as the increase to some would be greater than the decrease to others, economic welfare has risen. The increase to the winners has to be sufficient to compensate the losers, so that the losers would agree to the change. But compensation is not necessary for the conclusion that welfare has increased. It is thus possible that a change (such as a development project) increases the welfare of some rich people and decreases the welfare of some poor people: but as long as the rich gain more than the poor, there has been an increase in welfare.

It is sometimes argued that public policy should start by trying to achieve maximum economic welfare, irrespective of the distributional consequences. If politicians consider that the distribution of welfare is socially unsatisfactory, this should be tackled separately and independently, through the tax system (Shavell 1994).[19] It can easily be seen that this could lead to irresponsible public policy: each policy sector (such as spatial planning, or medical care, or transport) feels entitled to ignore the distributional effects, saying: "That is not our concern, but the concern of the fiscal service."

Main Conclusions

1. Spatial planning affects the use of economic resources, and in that way affects economic welfare.
2. Measuring economic welfare is not easy, but sometimes measuring the market outcomes of the spatial planning actions gives an approximation.

3. When applied to the economic evaluation of planning projects, this is the technique of cost–benefit analysis.
4. When applied to the economic evaluation of particular (legal) approaches to spatial planning, theories from welfare economics are required. Those theories usually start from the assumption of a perfectly free market. Then the effects for economic welfare are predicted when spatial planning is practised by 'structuring' markets, and by 'regulating' markets. However, those theories are not so well established that their predictions are fully reliable.
5. There are important theoretical and political objections to trying to expand an evaluation of the effects on economic welfare into a method for the *comprehensive* evaluation of spatial planning. In particular, the concept of economic welfare takes no account of the distribution of welfare between people, and it cannot take account of any moral issues (such as human rights, or sustainability) which might be affected.

Notes

1   Note that, according to these definitions, it is material goods and immaterial services which produce economic welfare. When *property rights* are subjected to an economic analysis –see also Chapter 2 – that which produces economic welfare is not goods and services, but *rights* in goods and services (Alchian & Demsetz 1973). This distinction should be noticed, but is of little importance at the academic level of this book.
2   The terms 'private goods' and 'public goods' are in common use but confusing, because they suggest that the distinction is between those who who should provide them and those who should own them: private persons for private goods and public agencies for public goods. However, the relationship is not so simple: a town hall, for example, is provided and owned by a state (public) body, but it is nevertheless a private good. See also Chapter 2.
3   Note that this classification of types of goods is not directly compatible with the classification of property regimes in Chapter 2: *private goods* = private property regime; *club goods* = common, or shared, property regime; *public goods* = public property regime + regime of open-access resources which are not over-used; *open-access resources which are over-used* = regime of open-access resources which are over-used.
4   However, the prices that people say they would be willing to pay might be higher than what they would pay in practice, for in practice people have limited budgets.
5   There is a form of econometric analysis called 'hedonic price theory' by which the monetary value of some properties of land and buildings which are not usually traded in the market can be estimated. The theoretical basis was laid by Rosen (1974). Nowadays the technique is widely used.
6   There are econometric models according to which the price effects of changing the distribution of wealth and income can be estimated (see, e.g., Bayer et al. 2004).
7   Note that there is a difference between a financial evaluation and an economic evaluation, even when the economic evaluation measures the effects in terms of

money. A financial evaluation takes account only of actual exchanges of money, whereas an economic evaluation takes account of changes in the value of all (relevant) goods and services.

8 It must be recognised that the statement "a perfectly free market would produce an economic optimum" is an untestable hypothesis, for it is impossible to discover empirically when economic welfare could not be higher; that is, when the maximum has been reached. The statement is theoretically interesting, and if it were true it would be practically useful. But it is not wise to base public policy on one untestable theory.

9 And when that has been attempted, as being socially or politically desirable, the result is often flourishing illegal markets. See, e.g., Nunez Fernandez (2012) for how dwellings are traded in Cuba.

10 Note that in the title of this book – 'Planning, law and economics' – the words 'law and economics' refer not just to that branch of economics called 'law and economics' but more widely; namely, to the application of welfare economics to the application of both private and public law.

11 The use of the term 'the public interest' is subject to the same misinterpretation, for it is often the case that individuals can be identified within 'the public' whose interest is at stake, and that not all individuals are served by 'the public interest'.

12 But see Coase (1974), where he provides evidence that, in the past, lighthouses – always considered as the providers of public goods (non-excludable and non-rivalrous) par excellence – have been provided privately.

13 "Given that one of the Paretian optimum conditions cannot be met, then an optimum situation can be achieved only by departing from the other Paretian conditions. The optimum situation finally attained may be termed a second best optimum . . . It is important to note that in general nothing can be said about the direction of the magnitude of the secondary departures from optimum conditions made necessary by the original non-fulfilment of one condition" (Lipsey & Lancaster 1956).

14 But, adds Coase (1960: 119), this says nothing about the equity of the distribution of the costs and benefits.

15 For an example of a gaming simulation carried out in order to investigate the effects of changing property rights on market outcomes, see Geuting and Needham (2012).

16 In 1729, Jonathan Swift (1667–1745) wrote "A Modest Proposal For preventing the Children of Poor People From being a Burthen to Their Parents or Country, and For making them Beneficial to the Publick". Swift suggested that the impoverished Irish might ease their economic troubles by selling their children as food for rich gentlemen and ladies. It was satire. But if a cost–benefit analysis had been carried out, omitting moral considerations, the evaluation might have been that the measure would increase economic welfare.

17 Suppose that you are concerned about the conditions which the children now being born will face – let us say the conditions thirty years in the future. You want them to have the same chances as you have now. If you discount the future at 5 per cent, the situation thirty years in the future has a weight of 23 per cent, so when making decisions now, you count what will happen in thirty years at 23 per cent of its value now. It is sometimes said that individuals have a 'defective telescopic faculty'; that is, private persons do not give enough weight to what will happen in the future. So one should discount not at 5 per cent but at a 'social discount rate', let's say, 2 per cent.

Then, the present significance of what will happen in thirty years is 55 per cent of its value now. In neither case are you taking the future of those being born now as seriously as you are taking the present. 'Discounting the future' is not a morally acceptable way of taking the account of the interests of future generations.

18 However, an economic evaluation can sometimes make explicit who are the winners and the losers. And, sometimes, weights are given to the effects depending on who experiences them (e.g. heavier weights if they are experienced by poorer people; see, e.g., Teulings et al. 2003). Such weights are necessarily arbitrary.

19 Such redistribution would lead to a different allocation of welfare, which would nevertheless be a new maximum. This is called 'the second law of welfare economics' (Dutta 1994: 15 et seq.).

References

Alchian, A. A., Demsetz, H., 1973. The property rights paradigm. *Journal of Economic History*, 33(1), 16–27

Bayer, P., McMillan, R., Rueben, K., 2004. *An equilibrium model of sorting in an urban housing market*. NBER Working Paper no. 10865. Cambridge, MA: National Bureau of Economic Resarch,

Bromley, D. W., 1991. *Environment and economy: Property rights and public policy*. Cambridge, MA: Blackwell

Bromley, D. W., 2001. *Property regimes and institutional change*. Paper presented at a conference on property rights and institutional change, Frederica, Denmark, 19–21 September

Buitelaar, E., 2004. A transaction–cost analysis of the development process: A method for identifying transaction costs in different institutional arrangements. *Urban Studies*, 41(3), 2539–2553

Coase, R. H., 1960. The problem of social cost. In Coase, R. H., 1988, *The firm, the market and the law* (pp. 95–156). Chicago: University of Chicago Press

Coase, R. H., 1974. The lighthouse in economics. In Coase, R. H., 1988, *The firm, the market and the law* (pp. 187–214). Chicago: University of Chicago Press

Cooter, R., Ulen, T., 2004. *Law and economics*, 4th ed. Reading, MA: Addison-Wesley

Dolfsma, W., 2013. *Government failure, markets and rules*. Cheltenham, UK: Edward Elgar

Dutta, B., 1994. Introduction. In Dutta, B. (ed.), *Welfare economics* (pp. 1–27). Delhi: Oxford University Press

Furubotn, E. G., Richter, R., 1991. The new institutional economics: an assessment. In Furubotn, E. G., Richter, R. (eds.), *The new institutional economics* (pp. 1–32). Tübingen: J. C. B. Mohr (Paul Siebeck)

Geuting, E., Needham, B., 2012. Exploring the effects of property rights using game simulation. In Hartmann, T., Needham, B. (eds.), *Planning by law and property rights reconsidered* (pp. 37–54). Farnham: Ashgate

Grant, M., 1988. *Forty years of planning control: The case for the defence*. Denman Lecture series. Cambridge, UK: Granta Editions

Hardin, G., 1968. The tragedy of the commons. *Science*, 162, 1243–1248

Harrison, A. J., 1977. *Economics and land use planning*. London: Croom Helm

Heikkila, E. J., 2000. *The economics of planning*. New Brunswick, NJ: Centre for Urban Policy Research

Lindblom, C. E., 2001. *The market system*. New Haven: Yale University Press

Lipsey, R. G., Lancaster, K., 1956. The general theory of the second best. *Review of Economic Studies*, 24, 11–32

Little, I. M. D., 2002. *A critique of welfare economics: A re-issue*. Oxford: Oxford University Press

Nelson, R. H., 2004. Local government as private property: towards the post-modern municipality, in Jacobs, H. M. (ed.), *Private property in the 21st century: The future of an American ideal* (pp. 95–124). Cheltenham, UK: Edward Elgar

Nunez Fernandez, A. R., 2012. *Urban land management in Cuba*. Utrecht: Uitgeverij Digitalis

Ogus, A. I., 1994. *Regulation: Legal form and economic theory*. Oxford: Clarendon Press

Ostrom, E. (ed.). 2007. *Understanding knowledge as a commons: From theory to practice*. Cambridge, MA: MIT Press

Pearce, B. J., 1981. Property rights vs. development control. *Town Planning Review* 52(1), 49–60

Pejovich, S., 1997. *The economic foundations of property rights*. Cheltenham, UK: Edward Elgar

Pennington, M., 2002. *Liberating the land: The case for private land-use planning*. Hobart Papers no. 143. London: Institute of Economic Affairs

Pigou, A. C. 1932. *The economics of welfare*. London: Macmillan

Rosen, S., 1974. Hedonic prices and implicit markets: Product differentiation in pure competition. *Journal of Political Economy*, 82(1), 34–55

Shavell, S., 1994. Why the legal system is less efficient than the income tax in redistributing income. *Journal of Legal Studies*, 23, 667–681

Stigler, G. J., 1992. Law or economics? *Journal of Law and Economics*, 35, 455–468

Teulings, C. N., Bovenberg, A. L., van Dalen, H. P., 2003. *De calculus van het publieke belang*. The Hague: Kenniscentrum voor Ordeningsvraagstukken, Ministerie van Economische Zaken

Tullock, G., 2008. Public choice. *The new Palgrave dictionary of economics*. London: Palgrave Macmillan

Webster, C., 2007. Property rights, public space and urban design. *Town Planning Review*, 78(1), 81–102

Webster, C., Lai, L. W. C., 2003. *Property rights, Planning and markets*. Cheltenham, UK: Edward Elgar

Williamson, O. E., 1999. Public and private bureaucracies: a transaction cost economics perspective. *Journal of Law, Economics and Organization*, 15(1), 306–342

Wolf, C., 1979. A theory of nonmarket failure: Framework for implementation analysis. *Journal of Law and Economics*, 22, 107–139

7

Law and Justice in Spatial Planning

What This Chapter Is About[1]

Spatial planning affects the formal rights which people can exercise over land and buildings, and also the access to and use of land and buildings by people who have no formal rights over them. Those are 'benefits' which land and buildings can bestow or take away, and spatial planning can affect the distribution of such benefits. Distributive justice is about the moral rightness of how benefits are distributed between people. Four different ideas about distributive justice are relevant here: utilitarian justice, egalitarian justice, justice as sufficiency, justice of frameworks. Each has its own implications for planning practice and for legal approaches to planning.

The Importance of Justice for Spatial Planning

Each and every spatial planning decision is related to issues of justice. Moreover, the main reason why modern spatial planning was introduced was to combat the perceived injustices of the terrible living conditions of the working poor in the late nineteenth-century cities (Hall 2014). Today, too, spatial planning measures are deliberately and explicitly targeted at injustices, or touch upon issues of (in)justice in more implicit and unintentional ways. Local governments, for instance, sometimes pursue social-mixing policies to stimulate more equal distributions of different income and ethnic groups over city neighbourhoods. At the same time, these policies are being accused of causing forced relocations and social exclusion of the poor and of ethnic minorities (e.g. Uitermark et al. 2007). The issue of justice is omnipresent, often through different and conflicting concepts.

This chapter is about the 'distributive justice' of rights and interests in land, and about spatial planning that can affect the distribution of those rights and interests. Justice is the principle of moral rightness. Distributive justice is the principle of moral rightness with regard to distribution (Buitelaar et al. 2017). Justice and fairness are often used interchangeably in common discourse.² What is a just or fair distribution of property rights over land? And what is a fair allocation of land uses? The concept of justice can then be applied to the:

–   distribution of the ownership and the value of the property rights
–   access to amenities which land and buildings can give (e.g. accessibility to work, shops, school, countryside, local amenity, environmental quality).

The question is asked here: how does the legal approach taken to planning affect that distribution and access?

The issue of justice may seem like an abstract, philosophical discussion – and indeed, there are many philosophical treatments of it. Nevertheless, it is of great practical importance, as the examples just given have demonstrated. Many day-to-day spatial planning decisions are made on the basis of an explicit or implicit concept of distributive justice. Planning authorities plan for affordable housing, as it is deemed 'just' that everyone should be able to occupy a decent home. And local governments provide libraries not only in affluent neighbourhoods, but also in neighbourhoods where people cannot afford to buy books. Decisions about a land-use plan often have considerable consequences for justice or fairness. Any designation of a building area discriminates between locations and thus its owners. Some landowners get richer (i.e. are able to realise valuable land uses), some see the value of their right decrease and adjacent landowners outside the building zone remain in their original position. In other words, spatial planning "makes people poorer or richer" (Needham 2006: 3). So, notions of justice are inherent and inevitable in spatial planning. The following section elaborates on this relationship between property rights, planning and justice.

Property Rights, Spatial Planning and the Distribution of Benefits

Property Rights and the Distribution of Benefits

Property rights lay down the conditions under which people can use and exchange land and buildings (see Chapter 2). Those rights are protected by law. In that way, existing distributions are protected, most importantly by making property rights exclusive. In the case of landed property, the fact that property rights are exclusive and the fact that land is immobile (i.e. attached to the earth's surface) creates (small) 'locational monopolies' in the sense that the titleholder(s) have a monopoly on the location to which the title applies. So ownership of rights inherently leads to unequal distributions: the title holder has everything (within the title), and all others have no claim on that land. Only those with property rights in homes in Manhattan or in Kensington can live in those areas; all others are excluded from doing that. This, of course, does not mean that these home occupants can use their rights unconstrained. Property rights are never absolute.

Spatial Planning and the Distribution of Benefits

Spatial planning affects property rights and therefore the distribution of ownership and of access to amenities. Spatial planning is locationally specific; it affects people in different locations differently. Sometimes those differences can have big financial effects. Spatial planning can redistribute the benefits to be obtained from land and buildings.

Often, spatial planning laws make it possible to compensate for *negative* financial effects, or even require this (see Alterman 2010, for international differences in the existence and size of 'planning compensation'). This may be the case when a planning authority changes the land-use rules on a particular plot owned by private actors, or when it provides infrastructure or buildings on its own, public land, in a way that reduces the value of the land and buildings of other title holders (see also Chapter 3). In some systems, the reverse is (also) possible: planning authorities may, through taxes and levies, take away *positive* financial effects caused by spatial planning (e.g. 'creaming off betterment', imposing conditions in return for 'planning gains', 'value capturing'). In those ways, spatial planning laws try to protect the *status quo ante*, the first distribution, by trying to prevent undeserved value increments and losses.

Nevertheless, particular acts of spatial planning often, and unavoidably, change that distribution, without the necessity to compensate (fully or partly) for the change (e.g. zoning agricultural land for urban development in some areas and not in others, or choosing to redevelop some housing areas and not others). Citizens often use their planning rights (Chapter 4) to try to avoid or mitigate losing some of the benefits they gain from the physical environment (e.g. NIMBY – 'not in my backyard') even though they hold no formal property rights in it.

When considering the inevitable issue of justice in spatial planning, it is important to recognise that there are different concepts of justice (Sandel 2007), with different and sometimes competing and contradicting ideas as to what is just and what unjust (Hartmann 2018). The assessment of a situation in terms of justice thus depends on which concept of justice is used.

In the following sections, four important concepts of justice are explained and related to spatial planning: utilitarian justice, egalitarian justice, justice as sufficiency, and justice as framework.

Utilitarian Justice

Utilitarian approaches look at utility or 'preference satisfaction'. Welfare economics is such a utilitarian approach. In the case of welfare, it is the preference satisfaction concerning scarce resources that is at stake (Pigou 1920). Chapter 6 discusses this in greater detail. Increasingly, the idea of preference satisfaction is extended to non-scarce 'resources' too. As a result, 'wellbeing', 'happiness' or 'broad welfare' are being considered as the overall indicator for standard of living (e.g. Layard 2005).

Although the focus on preference satisfaction (whether related to scarce or non-scarce resources) varies between different approaches, all those approaches are utilitarian in the sense that they aggregate individual utility and preference satisfaction. The resource allocation that leads to 'the greatest happiness for the greatest number' is to be preferred, according to utilitarian thinking (see, for instance, the pioneering work of Jeremy Bentham, 1748–1832; especially Bentham 1907). In other words, the distribution of property rights, and the access to amenities, that create the greatest overall utility or welfare, is to be preferred.

Overall welfare or overall happiness are relevant criteria for, for instance, evaluating various policy alternatives, and they have been worked out in practicable, advanced evaluation techniques such as cost–benefit analysis. However, they have their shortcomings. Chapter 6 discusses those more extensively, but there are three criticisms that are relevant in the context of (distributive) justice:

– Overall welfare and utility do not take account of the skewness of the distribution (e.g. the extent of the inequality in ownership of land and property).
– Little attention is paid to legally established (minority) rights, such as property rights and citizens' rights.
– Calculating very different preferences and values through one common currency – utility – is a problem.

Those criticisms get more attention in other perspectives on distributive justice.

Implications for the Legal Approach to Spatial Planning

Spatial planning from a utilitarian perspective should contribute to maximising happiness or welfare. Using welfare economics, Cost Benefit Analysis has been developed to predict *ex ante* which policy or project alternative will generate the greatest welfare. Spatial planning can increase utility by allowing for or providing goods that are valued as highly as possible.

Where markets alone will not achieve this, markets can be 'structured' and/or 'regulated'. Then, spatial planning can add to aggregate welfare by taking account of resources that are not priced but nevertheless valued. This includes regulating and mitigating negative externalities such noise nuisance, air pollution, poor amenity. And it includes stimulating positive externalities such as creating agglomeration effects by urban clustering, protecting and improving heritage, and allowing consumption- and production-generating uses such as department stores and head offices. It also includes actively providing goods that the market does not, or only insufficiently, supply, because they are non-excludable – such as infrastructure, public space, water and green space (see also Chapter 6). But this says nothing about the distribution of the utility to be got from the spatial planning.

Egalitarian Justice

Utilitarian justice takes no account of the skewness of distributions, unless utility is calculated separately for different distributions. The perspective of egalitarian justice *does*, however, take issue with skewed distributions. In the case of egalitarian justice, equality and justice are strongly related to each other. This can be in the form of "the more equal, the more just" (Smith 1994: 119), or that inequality should not be too great. Marxist perspectives on social justice belong within this category of justice (Harvey 1973).

Equality may concern all people here and now, but it might also be stretched to include non-human species or humans in future times (e.g. Nussbaum 2011). The latter is the issue of 'intergenerational justice'. Sustainability is the concept often used to assess the balance between current (ecological) interests and those of future generations.

When talking about (in)equality, it is important to be explicit about (in) equality of what. In the literature, a common distinction is between (in) equality of *outcomes* (e.g. access to housing) and (in)equalities of *opportunities* (and rights) (Buitelaar et al. 2017). The distinction is important because the two do not always coincide and are often mutually exclusive.[3]

Equality of outcomes relates to the *product* of social action, often measured in monetary terms such as income from labour and capital. Many are concerned with skewed income and capital distributions, as they are deemed to produce negative economic and social effects (e.g. Stiglitz 2012; Piketty 2014; Atkinson 2015) or to be, simply, morally undesirable. Although the discussion often focuses on labour and capital, the third factor of production is relevant here as well: land. Inequality in the possession of property rights over land and building, or access to amenities, may be the concern of public policy. On a large scale that policy can be through land reform programmes; on a smaller scale through spatial planning. For example, some people have (much) better access to green space or other services than others. In some cases, the difference may be considered too large, leading to policy action.

Equality of opportunities is about the differences between people in the options which they have to produce the same outcome. Not everyone has the same access to schooling or to the labour market, for instance, due to discrimination on the basis of ethnicity, class, age or sex. As a result, differences in income and capital generation and accumulation may occur.

One necessary condition in the quest to achieve equality of opportunity is equality of rights. This is not necessarily the same as equal possession of property rights; it concerns the rights to obtain these property rights (e.g. Moroni 2017). And it regards the equality of 'planning rights' or 'citizens' rights'. For instance, someone may not be excluded from acquiring property rights over land on the basis of the colour of his/her skin. And everyone who lives in a particular area has the (same) right to be heard by the planning authority and to appeal against the decision made by that planning authority. Such rights must be indiscriminate and impartial.

Equality of outcomes and equality of opportunity are related in the sense that they may reinforce each other – for instance, differences in possession of capital and land may create differences in opportunity. However, the two are not to be equated; they may even be at odds with each other. If you grant people equal opportunities through equal rights, the outcome is likely to be unequal because of differences in human nature (i.e. people have different talents) and differences in effort. Take, for instance, the 100 metre sprint: everyone is given the opportunity to start at the same point, but they all end in different times. Usain Bolt was clearly more gifted than his rivals. Conversely, creating equality of outcomes through redistribution requires treating people differently (i.e. not giving everybody equality of opportunities/rights), as it would require stimulating some and holding back or taking away from others (Nozick 1974).

There is another difference between outcomes and opportunities/rights that is relevant here. General *rights* are not scarce, whereas outcomes such as specific land and capital titles are. Someone can be granted the right to buy a house in the housing market without infringing upon the rights of others. Everyone may have that right, and adding one right does not come at the expense of another. However, in the case of land titles and capital (titles), for instance, there may a 'zero-sum' or even a 'negative-sum game', at least to some extent: one gains at the expense of another. Every person who wants to live within the Boulevard Périphérique of Paris makes it more difficult and costly for others after them to do the same. It is exactly because of this difference that the (in)equality of outcomes evokes such strong and polarised debates, and the equality of opportunities and rights receives almost universal support (at least, in words).

Implications for the Legal Approach to Spatial Planning

Spatial planning may contribute to the equalization of outcomes, such as property ownership and use, and the access to amenities. In the Netherlands, for instance, the spatial equality of service and amenity provision has been one of the longstanding priorities in the spatial planning culture (Hajer & Zonneveld 2000). The aim is that public, and to some extent private, amenities such as schools, hospitals, access to green space, public transport and shops be provided as evenly as possible. The equalisation of property ownership and use may occur, albeit incompletely, through planning compensation, value capturing and property taxes.

At a more general scale, equalisation of rights and opportunities is something that needs to be established primarily at the (pre-)constitutional and the legislative level. It means, for instance, prohibiting the practice of 'redlining': banks not providing mortgages to people living in particular (stigmatised and poor) neighbourhoods (Aalbers 2005).

Nevertheless, the administrative level may play a role in enforcing these rights and providing *de facto* rights and opportunities. It can, for instance, make sure that each neighbourhood has a more or less equal connection to the urban transport network in order to be able to get to the labour market. The provision of (affordable) housing is another example of creating *de facto* equal rights and opportunities alongside the *de jure* equal right to housing, since the latter is rather worthless if there is not enough (affordable) housing to effectuate that right.

Justice as Sufficiency

Egalitarian justice focuses on reducing or eliminating *relative poverty*: economic differences between people should be reduced, possibly eliminated.

This can include reducing (overly large) differences in land ownership or access to amenities. This view can, however, be criticised for its focus on differences rather than on people's (absolute) standard of living:

> The fact that some people have a lower standard of living than others is certainly proof of inequality, but by itself it cannot be a proof of poverty unless we know something more about the standard of living that these people do in fact enjoy.
>
> (Sen 1983: 159)

For instance, person A owns a house on ten hectares of land and person B on five hectares. There is clearly inequality: B owns and lives on twice as much land as A. However, A has more than sufficient to live on. Most people own or have access to much smaller plots. Furthermore, comparing two homeless people, there is perfect equality (i.e. none of the two occupies a home). Nevertheless, most would agree that there is a social problem: a home is a necessary condition for living a decent life – it is part of the minimum living standard.

The above shows that there can be great inequality with great prosperity and great equality with great poverty. Therefore, some argue for a focus on *absolute poverty* rather than relative poverty. That means not focusing on economic differences between people, but on whether people are living above or below a socially defined minimum standard:

> Economic inequality is not as such of particular moral importance. With respect to the distribution of economic assets, what is important from the point of view of morality is not that everyone should have *the same* but that each should have *enough*. If everyone had enough, it would be of no moral consequence whether some had more than others.
>
> (Frankfurt 1987: 21; emphasis in original)

This is known as the 'doctrine of sufficiency', as opposed to the "doctrine of egalitarianism" (Frankfurt 1987).

What 'enough' or the 'social minimum' is is determined by society and politics. It is therefore a contextual question, the answer to which varies over time and space. The threshold of decent or sufficient housing in a rich country is much higher than in developing countries. And, within (rich) countries, changes in this threshold over time can be seen. What many Western countries now consider to be the minimum standard for a home for poor people is of a much higher quality than that of homes in the slum neighbourhoods of the late nineteenth century (Hall 2014).

John Rawls (1971) refers to the concept of 'social primary goods', which is the minimum level of goods people should possess or have access to. A particular (to be contextually defined) income and wealth level is part of that. The concept can also be applied to space and spatial planning. Then 'spatial primary goods' can be identified: the minimum level of spatially connected goods that people should possess or have access to (Moroni 1997; Buitelaar et al. 2017).

Implications for the Legal Approach to Spatial Planning

Within a doctrine of sufficiency, spatial planning may play a role in making sure that people possess or have access to 'spatial primary goods'. Moroni (1997) distinguishes between four types of such goods: decent housing, accessible jobs, environmental safety and sufficient green space in close proximity. But their exact extent, nature and scope is, again, a contextual matter. The list can easily be extended (or limited) and further operationalised.

Spatial planning may set the rules to ensure the provision of spatial primary goods. For instance, many countries have building codes and ordinances that prescribe the minimal quality of a home and of buildings in general. Or local governments may require property developers to provide particular amenities and a certain share of affordable housing in return for planning permission, as happens in England through the use of 'section 106 agreements'. Instead of, or alongside, this more passive and permissive approach, spatial planning may play a role by actively providing spatial primary goods, such as infrastructure and green space.

Justice of Frameworks

In the previous three sections, three different perspectives (i.e. utilitarian justice, egalitarian justice and justice as sufficiency) have been discussed. They are all concerned, albeit in different ways and for different reasons, with the *distribution* of property rights, wealth, land, etc. However, there are also principles of justice that are not concerned with the distributional outcome, but 'only' with the justice of the *framework* within which that outcome comes about (e.g. Hayek 1982; Nozick 1974; Hayek 1982). In relation to spatial planning, "we can say the 'good city', the 'desirable city', cannot be defined in terms of certain features, but only in terms of the rightness of the framework of rules within which the city itself will emerge and function" (Moroni 2010: 147).[4] These are commonly referred to as liberal or libertarian views of justice.

An example of framework justice (rather than distributive justice) is Robert Nozick's 'entitlement theory of justice' (1974). It consists of two principles. The first is the principle of 'justice in acquisition', which refers to just *processes* of how 'unheld' things become held by people. This may, for instance, relate to how people assign property rights over land. The second principle is that of 'justice in transfer', which refers to the *processes* through which people exchange holdings, such as property rights over land. It concerns, on the one hand, legitimate means of exchange such as voluntary exchange, gifts, inheritance; and, on the other hand, illegitimate means such as exchange by force, fraud and monopolistic rent-seeking. According to Nozick, a distribution is just "if it arises from another distribution by legitimate means" (1974: 151). Legitimacy here means meeting the principles of 'justice in acquisition' and 'justice in transfer'. To make it concrete and relevant to spatial planning: if all people acquire land and buildings justly and exchange them justly (i.e. voluntarily), the distributional ownership pattern is just, regardless of the overall welfare it produces and regardless of how skewed the distribution is.[5]

Implications for the Legal Approach to Spatial Planning

In the case of justice of the framework, the role of spatial planning at the administrative level is limited, as changing and correcting distributional patterns by the imposition of public law rules is not just. The focus is on the justice of the framework of rules with regard to property rights and spatial planning (i.e. 'nomocratic planning'; see Chapter 3).

Part of that framework is property law, contract law and so on, which fall under private law. The establishment of private law is primarily a constitutional and legislative matter. Another part of the framework falls under public law. It is the establishment of rules for the compensation of value loss, or rules for capturing value increments of property rights, when these occur as a result of spatial planning measures. Planning compensation is paid in many countries if spatial planning limits the exercise of private property rights to an extent which is considered excessive. In the US, spatial planning can even go as far as to constitute a (regulatory) 'taking' (e.g. Alterman 2010). In such circumstances, compensation is a measure that is used not to achieve a particular distribution of property rights over a number of title holders, but to protect the institution of private property rights. It compensates for measures that infringe upon the right. Focusing on the justice of the rules of the game, regardless of their distributional output, is central to framework justice.

Planning compensation and value capturing are established in law, at the legislative level, but their implementation may be at the administration's discretion. In the Netherlands, for instance, property owners have a legal right to planning compensation (*planschade*) for spatial planning measures which decrease value – under certain conditions. However, paying for the unearned value increments that are the result of spatial planning decisions is not legally required of property owners. They are obliged to pay such a levy or tax (*baatbelasting*) only if the planning authority decides to capture that value – again under certain conditions.

Justice in Spatial Planning Practice

In practice, different perspectives on justice may coexist or compete. For instance, some governments try to pursue all four principles of justice at the same time, at least to some degree. Local governments, for instance, want a provision of amenities such as infrastructure and parks that boost overall local welfare (*utility*). At the same time, they want the differences in the access to these amenities to be *not too unequal* among the population, and they want there to be at least *enough* for all constituents. And finally, in liberal democracies, governments want people to be able to acquire and change their location (i.e. their proximity and access to amenities) justly, *voluntarily, without coercion*.

Although different principles may be pursued at the same time, this does not mean they are (entirely) compatible. They are often even antagonistic: more of one means less of the other. Take, for example, the relation between 'egalitarian justice' and 'justice of frameworks' in relation to urban segregation. If an important principle of a just framework is granting people the right to settle freely, the spatial distributional pattern that results from that is very likely to be one of segregation according to income, class, age and/or ethnicity. This is because of people's residential *preferences* – 'birds of a feather flock together', and different groups want different facilities which are themselves not evenly distributed – and because of *ability*, as not everyone can afford to live in every neighbourhood. Aiming for desegregation, or much less segregation, necessarily implies limiting the liberty of people to settle where they want and are financially able to. In other words, the two principles of justice ('egalitarian justice' and 'justice of frameworks') are then at odds with each other: there cannot be desegregation and settlement liberty at the same time.

Main Conclusions

1. Spatial planning affects the distribution between people of the bene-fits to be derived (directly or indirectly) from land and buildings. Judging the moral rights of that distribution, or of changes in it, is a matter for distributive justice.
2. Spatial planning can therefore always be evaluated by the standards of distributive justice.
3. There are four schools of thought about distributive justice which are applicable to spatial planning: utilitarian justice, egalitarian justice, justice as sufficiency, justice of frameworks.
4. Choosing for one or the other of those schools has important implica-tions, not only for the content of the spatial planning, but also for the way (the legal approach) in which it is carried out.
5. In practice, more than one of those principles can be applied in a particular planning policy, in which case none of them is applied in an extreme form.

Notes

1   The authors gratefully acknowledge the comments made by Professor Stefano Moroni on an earlier version of this chapter.
2   In political philosophy, 'fairness' is specifically linked to Rawls' concept of 'justice as fairness'. Here it is used in the way it is commonly used in social discourse.
3   Equal rights are commonly considered to be part of libertarian views on justice, as formulated by Thomas Jefferson, Milton Friedman, and others who put equal opportunities before equality of outcome (as promoted by John Rawls). Those two views can be placed under the broad concept of egalitarian justice. For a more elaborate distinction of the various forms of justice, see Sandel (2010).
4   In the literature, 'framework justice' is commonly referred to as 'nomocracy' (Hayek 1982; Moroni 2010).
5   According to Rawls, there are various problems with this. One is that Nozick is not concerned with how just the initial distribution or basic framework was or has come about. When one starts trading from an unjust starting point, such as not everyone possessing basic 'social primary goods' (see 'Justice as Sufficiency'), every individual voluntary transaction may be just, but the structure itself and the final distribution ceases to be just (Rawls 1993: 266). Another problem can arise with the rules for 'acquisition' and for 'transfer'. Who sets these, and who determines whether or not they are just?

References

Aalbers, M. B., 2005. Place-based social exclusion: Redlining in the Netherlands. *Area*, 37(1), 100–109

Alterman, R., 2010. *Takings international: A comparative perspective on land use regulations and compensation rights*. Chicago, IL: ABA Press

Atkinson, A. B., 2015. *Inequality – what can be done?* Cambridge, MA: Harvard University Press

Bentham, J., 1907 (2007). *An introduction to the principles of morals and legislation.* Mineola, NY: Dover Publications

Buitelaar, E., Weterings, A., Ponds, R., 2017. *Cities, economic inequality and justice: Reflections and alternative perspectives.* Abingdon, UK: Routledge

Frankfurt, H., 1987. Equality as a moral ideal. *Ethics,* 98(1), 21–43

Hajer, M., Zonneveld, W., 2000. Spatial planning in the network society – rethinking the principles of planning in the Netherlands. *European Planning Studies,* 8(3), 337–355

Hall, P., 2014. *Cities of tomorrow: An intellectual history of urban planning and design since 1880.* Oxford: Wiley-Blackwell

Hartmann, T., 2018. Ethik in der Raumplanung. In Blotevogel, H. H., et al. (eds.), *Handwörterbuch der Stadt- und Raumentwicklung.* Hannover: ARL, in press

Harvey, D., 1973 (2009). *Social justice and the city.* Baltimore, MD: Johns Hopkins University Press

Hayek, F. A., 1982 (2013). *Law, legislation and liberty.* London: Routledge

Layard, R., 2005. *Happiness: Lessons from a new science.* London: Penguin

Moroni, S., 1997. *Etica e territorio.* Milan: Franco Angeli

Moroni, S., 2010. Rethinking the theory and practice of land-use regulation: Towards nomocracy. *Planning Theory,* 9(2), 137–155

Moroni, S., 2017. Property as a human right and property as a special title: Rediscussing private ownership of land. *Land Use Policy,* 7, 273–280

Needham, B., 2006. *Planning, law, and economics: The rules we make for using land,* 1st ed. Abingdon, UK: Routledge

Nozick, R., 1974. *Anarchy, state and utopia.* New York: Basic Books

Nussbaum, M., 2011. *Creating capabilities: The human development approach.* Cambridge, MA: Belknap Press

Pigou, A. C., 1920. *The economics of welfare.* London: Macmillan

Piketty, T., 2014. *Capital in the twenty-first century.* Cambridge, MA: Harvard University Press

Rawls, J., 1971. *A theory of justice.* Cambridge, MA: Belknap Press of Harvard University Press

Rawls, J., 1993. *Political liberalism.* New York: Columbia University Press

Sandel, M. J., 2010. *Justice: What's the right thing to do?,* 1st ed. New York: Farrar, Straus and Giroux

Sen, A., 1983. Poor, relatively speaking. *Oxford Economic Papers,* 35(2), 153–169

Smith, D. M., 1994. *Geography and social justice.* Oxford, UK: Blackwell

Stiglitz, J. E., 2012. *The price of inequality: How today's divided society endangers our future.* New York: W. W. Norton & Company

Uitermark, J., Duyvendak, J. W., Kleinhans, R., 2007. Gentrification as a governmental strategy: Social control and social cohesion in Hoogvliet, Rotterdam. *Urban Studies,* 39, 125–141

8
Law and Legitimacy in Spatial Planning

What This Chapter Is About

A state body carrying out spatial planning can do that more quickly and more effectively if its actions are regarded as legitimate by its citizens. Otherwise, its actions – even if they are formally legal – will be met with distrust and suspicion, leading to opposition. Legitimacy can be earned in three ways: if the state body is recognised by the public as acting on their behalf (input); if the body follows the procedures prescribed for protecting citizens' rights (throughput); and if the body produces results which are generally recognised as good (output). However, a planning authority which earns legitimacy in one of those ways might not be seen as legitimate in one or more of the other ways. Rules of good governance can help a planning authority to choose a legal approach which its citizens will recognise as legitimate.

What Is Legitimacy?

Spatial planning is the activity of a government (state) body. The public interest that the governmental body pursues can be in conflict with private landowners' interests, and the spatial planning affects the distribution of costs and burdens of land uses. Therefore, planning authorities are equipped with special powers to realise their planning aims – if necessary against the wishes of some of the private interests. The planning instruments use public power, trust and money to influence how people may exercise their property rights (and see Chapter 3). Planning is thus a governmental action that intervenes in how people exercise their rights. In a democratic society, peaceful compliance to such interventions requires some general acceptance of the governmental action in question. The content and degree of this public acceptance is called legitimacy (Bekkers 2007; Mees et al. 2014).

Legitimacy and Citizens

The framework within which legitimacy is acquired depends on the way the relationship between the citizen and the state is constituted, in particular by the form of democracy. There are different forms of democracy, such as representative, deliberative or direct democracy (Bekkers 2007), and with each form, specific institutions (laws, public bodies, etc.) contribute to legitimacy. Switzerland, for example, is an example of a direct democracy, whereby many decisions are made through citizen initiatives and referenda. Germany is an example of a representative democracy, in which citizens delegate their decision-making power to representatives, who decide on their behalf.

In a wider perspective, legitimacy is related to the trust between the governing and the governed. Citizens might be more willing to accept public actions if they trust that the action in question serves the public interest and if they trust the state body to act honestly. The importance of trust for legitimacy has two implications. The first is that it can take a long time to gain such trust, but it can be lost quickly.[1] Second, the trust between citizens and the state depends on specific historically contingent conditions. It has been argued, for example, that the Dutch polder model, which is based on mutual trust between the state, citizens and non-governmental organisations, has arisen because of the necessity for united action in the fight against flooding (Schreuder 2001).

Inherent to the concept of legitimacy is the idea that the state is answerable to its citizens in all its actions. It must be able to justify its actions. And if it does that by arguing that those actions are in the public interest, the state body must be prepared to let the citizens influence those actions, directly or indirectly. This influence may be very direct (such as in direct democracies) or indirect (such as in representative democracies) or it may be required as part of the planning process (giving people the right to contribute in one way or another; see Chapter 4). The law institutionalises this influence, although the way of legitimation is rarely explicitly stated in the text of the law.

Legitimacy – A Political Concept

Whether a particular action of a state body is legitimate, or regarded by the citizens as legitimate, cannot be determined or measured objectively. Rather, legitimacy is subject to continuous political and academic reflection. For example, when new challenges of spatial planning emerge, such as climate change or other environmental or socioeconomic changes, there is a need to reflect on the way spatial planning measures can be legitimised (van Buuren et al. 2012; Hartmann & Spit 2016). The concept of legitimacy is inherently political.

This can be illustrated with the following example. Assume a municipality wants to develop a certain area of a city, and it does this by first acquiring the land where the development is to take place (here called 'an active land policy'). To facilitate this, the municipality joins in a public–private partnership with a commercial real estate developer. The municipality will use public power, trust and money to acquire the land – for example, by applying preemption rights or compulsory purchase. The profits of the development are shared between the private developer and the municipality. This can result in the private developer profiting directly from the public intervention (i.e. the land acquisition). In the Netherlands, this is generally accepted as a fully legitimate form of public spatial planning; in other countries it might be regarded as illegitimate, and is sometimes even illegal.

Legality and Legitimacy

Legality is a related concept: it describes whether some action (in this book, a planning action by a state body) is in accordance with the law. Most of the time, legality and legitimacy are closely related. Most actions that are legal are also legitimate, and vice versa. But the fact that legitimacy is a quality that must be earned implies that it can change; legality, in contrast, is more stable and robust. Consequently, tensions can arise between legitimacy and legality.

In particular when the situation is changing rapidly or radically, the legitimacy of actions can be questioned, even if those actions are within the scope of the law (i.e. legal). Examples are environmental changes that affect property rights (van Straalen et al. 2018), and socioeconomic dynamics (Hartmann & Needham 2012). The close collaboration between a state body and a private real estate developer, described above as being very common in the Netherlands, might not be against the planning law (or other laws) – it is thus legal. But after the real estate crisis in 2008, the legitimacy of many forms of that 'active land policy' is being questioned. The public acceptance of this practice can no longer be taken for granted. Another recent change is rivers flooding their banks: this has affected the legitimacy of allowing building in floodplains. A municipality in Germany approved a land-use plan for building in a potential flood area. The mayor argued that this plan was legally approved by the regional planning authority (Hartmann 2011). However, the public acceptance for such developments diminished after a series of flood events. So, the legality of the plan remained, but the legitimacy was questioned. These examples illustrate how external circumstances can affect the legitimacy of an action, even though it remains legal. And it can even be so that an illegal action of a planning authority is silently tolerated, because it is accepted as effective and efficient, not harmful, and therefore legitimate.

Forms of Legitimacy: Input, Throughput, Output

How can governmental actions be legitimised? Three major 'forms of legitimacy' have been distinguished: input, throughput and output legitimacy (Scharpf 1999; Schmidt 2013). Input legitimacy is derived from the agency that takes the action, throughput legitimacy is derived from the process by which the action is taken, and output legitimacy is derived from the result that is achieved.

Input Legitimacy

Input legitimacy refers to the quality of representation of the public interest in the state body taking the action. The normative idea behind this is that of 'government by the people'.[2] In other words, if the state body has been elected in a democratic way, it is empowered to act on behalf of the citizens. No additional consultation about specific issues is necessary, and the state body does not need to justify an action by the achieved result. Governmental actions are legitimate because they are enacted by representative, authorised and accountable organisations.

Imagine a police officer arresting a criminal. This action does not require public consultation, nor is the action of the policy officer assessed against the achieved result. Rather, the police officer may arrest a criminal because of the delegated power received via the institutional system.

There are few examples of pure input legitimacy in land-use planning (handling a building application might be such an example), but they are more common in sectoral planning. For example, in some countries water management is carried out with a strong input legitimacy. In the Netherlands, the Directorate-General on Water (*Rijkswaterstaat*) is the central body for water management (Hartmann & Spit 2016) – it governs on water-related issues with a 'hegemony of the state' (Wiering & Crabbé 2006). Although there is a trend towards other forms of governance in this sector, many spatial measures for water management – such as dyke projects – are still legitimised through input legitimacy. These measures can substantially influence spatial planning and other policy sectors, which might operate under one of the other legitimacy regimes.

Throughput Legitimacy

Throughput legitimacy results from the quality of the decision-making processes. Governmental actions are thus legitimate if citizens and relevant stakeholders are enabled to contribute to an effective, accountable and transparent process (Schmidt 2013). Throughput legitimacy is therefore strongly linked with citizens' rights (Chapter 4).

Throughput legitimacy is based on all (relevant) actors being able to interact. If 'due process' has been followed (see Chapter 4), the action is thereby legitimate. Participatory and collaborative practices introduced in the last few decades contribute to this form of legitimacy (van Coenen et al. 2001; Mickel et al. 2005). However, meaningful or effective participation (in the sense that the participation influences the content of the decisions) is not necessary for throughput legitimacy: as long as the procedural steps have been followed properly, the action is legitimate.

Throughput legitimacy is common in spatial planning. For example, the legal procedure for making and approving a binding land-use plan often prescribes certain obligatory steps, such as participation in an early stage, stakeholder involvement, approvals by certain governmental bodies (e.g. to check the compliance with regional planning or sectoral plans), public display, etc. (also see Chapters 3 and 4). The procedural steps legitimate the plan. If the procedure has not been properly followed, this can give grounds for objecting against the plan. Stakeholders who object to the content of a spatial plan might be able to attack it in that way; that is, procedurally. For that reason, it is prudent for the planning authority to follow the prescribed legal procedures.

One of the challenges of throughput legitimacy is to determine who the relevant stakeholders and citizens are and how to involve them. For many small projects this is relatively easy, but for projects that have a large spatial impact this is particularly difficult. Imagine that a station area for a large city is to be redeveloped. Who should play a role in the process: only the inhabitants of that city, or commuters who travel through the area every day, as well?

Output Legitimacy

Output legitimacy is achieved when state actions realise results which most people support. If people are pragmatic and say 'The ends are so important that they justify the means', then no further legitimation is necessary. Output legitimacy works best in situations where the goals and values at stake are highly consensual; where an action is clearly in the public interest.

The realisation of a certain number of workplaces, provision of social housing, solution of a traffic problem – these can be the desired results of spatial planning projects. However, a particular plan can aim to tackle many different problems (there are many different outputs). If it solves just some of them, leaving others unsolved, that is not sufficient to legitimise the whole plan via output legitimacy.

Output legitimacy is closely related to effectiveness, as discussed in Chapter 5. This means that, if an action is to be legitimised by its output, that legitimacy can be increased by increasing its effectiveness in achieving the adopted aims.

Choosing a Legal Approach

Legitimacy is one of the conditions which a planning authority takes into account when choosing a legal approach. How can it do that?

Sometimes, the law leaves little room for choice. For example, certain activities are assigned to the competency of named institutions, such as water authorities who are responsible for the construction and maintenance of flood defences. The actions of those authorities derive input legitimacy from the law. Sometimes the law requires that an action be justified by its output. Expropriation, for example, has to be legitimated by the result to be achieved: it is not justifiable if it is not in the public interest (leaving aside how broadly or narrowly this is defined and interpreted in different countries). If it is required that a planning project be supported by a cost–benefit analysis, this is a form of output legitimacy. Many planning actions are required to follow specified legal procedures; in that way, they acquire throughput legitimacy.

But the law often allows room for choice. This can be because the law is not binding or does not include all possibilities, for even in legal texts there can be editorial statements and paragraphs without explicit legal consequences. Then judicial analysis might be necessary.

Such an understanding can help a planning authority to choose a legal approach for its activities. For example, in cases where projects promise some measurable output (e.g. a housing development on a greenfield site at the edge of a growing city), the planning authority can concentrate on effectiveness: it justifies its actions by the output. If the outputs and planning objectives are vague – such as often is the case with brownfield developments – it might be better to rely on input legitimacy. If there is little consensus about the content of a project which is nevertheless supported by the authority, it might not be sensible to aim for output legitimacy. Then throughput legitimacy is necessary; the procedures must be followed carefully. Objections to the content will not go away, but they may legitimately be dismissed if the prescribed procedures have been carried out. Throughput legitimacy can also be important for complex inner-urban areas, where stakeholder roles are difficult to identify and are distributed among many parties, and the output is difficult to measure. If a planning action can be fully justified by input legitimacy, procedures and output need less justification.

For a more concrete illustration, take an urban development. There might be a choice between three legal approaches:

— an active land policy (i.e. actively acquiring land that is needed for a certain development and developing it)
— inviting a public tender for best ideas
— participatory approaches to involve local citizens and develop a bottom-up development strategy.

Each of those approaches is legitimised in different ways:

— The active land policy approach is justified by its output: the financial result and probable effectiveness.
— The public competitive tender relies on input legitimacy: the state body determines from its public role the terms of the tender and chooses (possibly assisted by a jury) the best proposal.
— The participatory approach and stakeholder involvement builds on throughput legitimacy.

Earning Legitimacy in Practice

For a planning authority to do its work well, its actions need to be recognised as legitimate. Suppose, however, that input legitimacy is satisfied but not throughput or output legitimacy. Or that the output is realised as promised but by a planning authority which is not fully recognised by the public, or by ignoring the prescribed procedures. One form of legitimacy alone might not be sufficient. This is illustrated below with a practice very common in Dutch planning.

The Double-Hat Problem of Dutch Active Land Policy

Spatial planning creates increases in land rent by assigning development rights to specific plots of land. The owner of the respective plot of land might receive a substantial increase in land values, called development gain. In active land policy, planning authorities can make use of this planning gain by acquiring land at its existing use value before they make the land-use plan public, and selling it at its higher value. In this way, the planning authority captures the development gain. This gain can then be used for realising public infrastructure, for building public facilities and for buying more land. The state body uses public power, money and trust to acquire the land, which can include the use of instruments such as expropriation (although rarely used) or municipal preemption rights (Needham 2014). The municipality acts thus as a real estate developer and a planning authority at the same time. It can be argued that it is justifiable to capture the development gain, because that is used for the public interest and also because it has been produced by the society as a whole, not by the individual landowner. (This concerns the justice of active land policy; see Chapter 7.)

However, the double role which the municipality plays can create issues of legitimacy, sometimes referred to as the 'double-hat problem'. It is a problem because, for many Dutch municipalities, active land policy has become a substantial source of income (Buitelaar 2010). This sets an incentive for the planning authority to develop locations in such a way that will bring a profit, possibly at the expense of non-profitable developments. Output legitimacy is derived from a successful development (and if the profits are used to subsidise development elsewhere, that also will earn output legitimacy). This can distort planning decisions, because of the lure of development gain. The danger is that the municipality puts its input legitimacy at risk: it wants the citizens to trust it to choose planning locations 'in the public interest', rather than the citizens regarding it as a commercial developer justifying itself by development profits. An additional danger of relying on output legitimacy in that way is that the outputs might be realised only after a long time, while input and throughput legitimacy have been questioned from the beginning.

Another legitimacy problem for the planning authority can arise because that state body can employ two sorts of powers in its interactions with private and commercial actors on the real estate market. The planning authority has the private law powers of any legal person (see Chapter 2) and also the public law powers entrusted to it as a planning authority (see Chapter 3). It can use those latter powers to grant or refuse planning permissions, even to (threaten to) use expropriation. Then it damages its throughput legitimation, as well as its input legitimation.

Good Governance and an Ethical Code for Planning Agencies

It will be clear that a state body which wants the citizens to accept its actions (including its spatial planning) as fully legitimate cannot ensure that by simply following a set of legal rules. Something more complex is needed and, in response to that, bodies of practice have arisen which can be summarised under the name of 'good governance' or 'principles of responsible government'.[3] Although these principles are not formally determined in legislation in every country, they provide a judicially recognised guideline for governmental activities (Needham 2014). They are described and discussed in some detail in Chapter 4 of this book.

Such principles can be very pragmatically formulated, or more systematically based on an ethical code for planning authorities. It is noteworthy that most attention for ethics in planning goes to the ethical principles which should guide the actions of individual planners. This book, however, emphasises the importance of an ethical code for planning authorities, because it is they, rather than the planners, who are empowered to act (Hoekveld & Needham 2013).

Main Conclusions

1. Any state body, including a planning authority, needs to take account of the legitimacy of its actions. That is more than legality; it is about the trust which the citizens have in the state body and the support they give to those public actions.
2. A planning authority can try to earn that legitimacy in three ways: by relying on its status as a representative of the people, by following prescribed procedures wholeheartedly and by producing results which (most) people fully support.
3. Each way is independent of the others, and the importance of each to the citizen can depend on the circumstances.
4. However, if one of the three is not followed, this can damage the legitimacy of the whole planning policy, and can lead to delays and legal procedures.

Notes

1 There is a Dutch saying: 'Trust comes on foot, disappears on horseback.'
2 This fits the idea of democracy where the state acts in the interest of citizens (van Buuren et al. 2012), which is theoretically based on societal contract theory such as *Leviathan* by Thomas Hobbes (Davy 1997; Scharpf 1997).
3 This is a translation of the Dutch '*beginselen van behoorlijk bestuur*'.

References

Bekkers, V. J. J. M., 2007. *Governance and the democratic deficit. Assessing the democratic legitimacy of governance practices*. Aldershot, UK, and Burlington, VT: Ashgate

Buitelaar, E., 2010. Cracks in the myth: Challenges to land policy in the Netherlands. *Tijdschrift voor Economische en Sociale Geografie*, 101(3), 349–356

Buuren, A. van, Klijn, E.-H., Edelenbos, J., 2012. Democratic legitimacy of new forms of Water Management in the Netherlands. *International Journal of Water Resources Development*, 28(4), 629–645

Coenen, F. H. J. M., Poppel, R. van de, Woltjer, J., 2001. De evolutie van inspraak in de Nederlandse planning. *Beleidswetenschap*, 15(4), 313–332

Davy, B., 1997. *Essential injustice. When legal institutions cannot resolve environmental and land use disputes*. Vienna and New York: Springer

Hartmann, T., 2011. *Clumsy floodplains: Responsive land policy for extreme floods*. Farnham, UK: Ashgate

Hartmann, T., Needham, B. (eds.), 2012. *Planning by law and property rights reconsidered*. Farnham, UK: Ashgate

Hartmann, T., Spit, T., 2016. Legitimizing differentiated flood protection levels – consequences of the European flood risk management plan. *Environmental Science & Policy*, 55, 361–367

Hoekveld, G., Needham B., 2013. Planning practice between ethics and the power game. *International Journal of Urban and Regional Research*, 37(5), 1638–1653

Mees, H. L. P., Driessen, P. P. J., Runhaar, H. A. C., 2014. Legitimate adaptive flood risk governance beyond the dikes. The cases of Hamburg, Helsinki and Rotterdam. *Regional Environmental Change*, 14(2), 671–682

Mickel, W. W., Bergmann, J. M., & Grupp, C. D., 2005. *Handlexikon der Europäischen Union*, 3rd ed. Baden-Baden: Nomos

Needham, B., 2014. *Dutch land use planning: The principles and the practice*. Farnham, UK: Ashgate

Scharpf, F. W., 1997. *Games real actors play: Actor-centred institutionalism in policy research*. Boulder, CO: Westview Press

Scharpf, F. W., 1999. *Governing in Europe: Effective and democratic?* Oxford and New York: Oxford University Press

Schmidt, V. A., 2013. Democracy and legitimacy in the European Union revisited: Input, output and "throughput". *Political Studies*, 61(1), 2–22

Schreuder, Y., 2001. The polder model in Dutch economic and environmental planning. *Bulletin of Science, Technology & Society*, 21(4), 237–245

Straalen, F. van, Hartmann, T., Sheehan, J. (eds.), 2018. *Property rights and climate change. Land-use under changing environmental conditions*. Abingdon, UK, and New York: Routledge

Wiering, M., Crabbé, A., 2006. The institutional dynamics of water management in the low countries. In Arts, B., Leroy, P. (eds.), *Institutional dynamics in environmental governance* (pp. 93–114). Dordrecht: Springer

9

Using the Law in Practice

What This Chapter Is About

This book has been written in order to show the importance of law for spatial planning, and the various ways in which practice can apply the relevant laws. This final chapter synthesises the arguments presented in the previous chapters, in order to derive conclusions about how a planning authority can choose a legal approach to its spatial planning. The book does not advise or prescribe which approach should be chosen (how the law should be used). Instead, it aims to help a planning authority to make that choice by identifying eight questions that body needs to answer. They can be regarded as variables; there are very many possible positions to be chosen within each one, and there are logical connections between them. The position a planning authority takes within each variable is influenced by social and economic considerations, as well as by the external context. In order to demonstrate how this framework of eight variables can be used in practice, the planning story started in Chapter 1 concludes here.

The Choices Leading to a Legal Approach

A planning authority, when determining its spatial planning policy for a particular location or set of problems, and how that policy is to be realised in practice, takes a position – explicitly or implicitly – within each of eight separate variables These are the following.

The necessary prelude for any planning measures:

–   What are the aims of the spatial planning for this location and/or for these problems, and how should they be formulated?

For each of the three types of relevant law described in this book:

- How to take account of the existing property rights in the planning area?
- How to use the available planning laws?
- How to take account of citizens' rights?

For each of the four types of considerations described in this book as being considerations that should influence the choice of a legal approach:

- What degree of effectiveness and efficiency is desired?
- How much importance to give to economic welfare?
- What interpretation of justice to use?
- How are the planning actions to be legitimized?

The answers the planning authority gives might vary according to the nature of the spatial plan or policy being pursued. And the position taken within each variable can have consequences for the position taken within the other variables. The answers given to all of those eight questions together lead to a choice of the legal approach that the authority takes to the particular plan or policy.

Below, each of the eight variables is discussed. It will be apparent that the position which can be taken within each of them is usually restricted.

What Are the Aims of the Spatial Planning for This Location and/or These Problems, and How Should They Be Formulated?

It might be that it is decided to create the conditions under which people can act and interact without causing each other harm or disturbance, or conditions which facilitate fruitful interactions between people – in both cases leaving people as free as possible to choose their actions. In that case, the planning authority has not imagined a particular land use it wants to see realised.

Or the planning authority might want to see realised a particular land use within part or all of its jurisdiction, specified in more or less detail. If that is so, those aims might be restricted by planning policies of other state bodies to which the planning authority is bound to a greater or lesser degree (vertical coordination). For example, there might be a landscape which is protected by national rules, or a route reserved for a new national motorway, which local planning authorities must respect. And a planning authority might be legally obliged to take account of its existing policies for other sectors (horizontal coordination).

How to Take Account of the Existing Property Rights in the Planning Area?

The position to be taken is limited by the existing property rights, by the rules protecting them and by the distribution of those rights between holders. A planning authority can nevertheless decide to try to acquire some of those property rights, either amicably or with varying degrees of legal force, up to compulsory purchase.

How to Use the Available Planning Laws?

Clearly, the planning authority is bound to observe the existing planning laws, but it might nevertheless have a wide range of freedom within them. In particular, it might be possible to lay down plans and policies with more or less detail; it might be possible to use the freedom granted for discretionary decisions; it might be possible to change the existing approved plans with more or less speed. Also, it will usually be possible to choose the size of the plan area: one plan for a big area, or many little plans.

How to Take Account of Citizens' Rights?

Here, also, the planning authority is bound to observe the existing laws which protect citizens' rights. Nevertheless, it can do that generously or grudgingly: by encouraging citizens to become involved, by following the minimum rules with reluctance or by ignoring those rules and hoping that no one will notice.

What Degree of Effectiveness and Efficiency Is Desired?

How effective does the planning authority want its plans and policies to be? This is partly a question about the level of detail with which it wants to determine the desired land use. The less the detail, the more likely the plan is to be effective. It is also a question of urgency: if there is a big problem which needs to be solved quickly (such as the imminent danger of flooding), then the plan should be highly effective.

Then there is the question of how efficient the spatial planning should be in realising its desired effects. A planning authority will want to use its available resources – expertise and budget – as efficiently as possible. This consideration is likely to be more relevant, the bigger the change in land use that is desired, for that usually requires more preparation, public infrastructure and guidance.

How Much Importance to Give to Economic Welfare?

The planning authority should take account of the effects of its spatial planning on the economic welfare of the citizens. Again, this consideration is likely to be more relevant, the bigger the change in land use that is envisaged, for that usually involves more economic resources in realising and using the new development.

What Interpretation of Justice to Use?

The planning authority needs to be aware of the interpretation which it gives to justice in spatial planning. That interpretation might differ according to the nature of the plan or policy. For example, a policy to increase sustainability implies a concern for intergenerational justice, whereas a policy to replace slum housing with better dwellings implies a concern for lifting some people out of socially unacceptable conditions (justice as sufficiency).

How Are the Planning Actions to Be Legitimised?

The planning authority needs to decide how important this consideration is for its own place in the society. If it already enjoys the acceptability of the public, it can perhaps rely on them to trust it to continue to act honestly and openly. If that acceptance is lacking, the authority might want to earn it by doing its best to prove its legitimacy. Or the planning authority might act autocratically, relying on its formal powers to push through its plans and policies and hoping that the results will be so good that the citizens will let the ends justify the means.

The National Government as Planning Authority and as Legislator

Note that the *national government* can be a planning authority in its own right. It can, in addition and in contrast to lower levels of government, be the legislator. (In some countries, the region or federal state might be the legislator.) As the legislator, it is not as restricted as the lower levels, because it can change the laws relating to property rights, spatial planning and citizens' rights. However, such laws are not changed easily or quickly. If the national government (or other legislator) pursues a particular planning policy, it has to obey the existing laws. It can at the same time decide to try to change some of those laws, to give different possibilities for its future planning policies.

The Interactions Between Those Eight Variables

There is a huge number of possible ways of combining possible choices within each of those variables – the choice of the planning aims, combined with the choice of what to do with property rights, combined with the choice of how to use the planning law, combined with the choice of how important effectiveness and efficiency are, etc. This book will not try to give an overview of those possibilities. Moreover, some of the combinations are logically impossible or not advisable because of the interactions between the choices. Below are some examples of those interactions.

If for a particular location the planning authority chooses to go for little change, and if the aims are formulated in general terms (e.g. it is desired to retain the agricultural character of a rural area), then the planning law can be used passively (no change without prior permission) and the available room for discretion used sparingly. It can be expected that the plan will be effective and realised efficiently (with little input), that economic welfare will change only slightly, that justice will be done in the sense that the existing distribution of land and buildings is largely retained and that – as long as that is done by following the rules strictly and openly – the planning authority will retain its legitimacy.

If the aims of the planning authority are for large-scale redevelopment of a town centre, and if the desired new town centre is specified in detail, this is unlikely to be realised without the planning authority acquiring some land and buildings. The collaboration of other property owners and of property developers is necessary and, in order to acquire that, some of the details in the plan might have to be changed. Effectiveness, in the sense that the plan is realised in detail, is then low. Efficiency too is not likely to be high, because of the big input required from the state body. If property is acquired amicably rather than compulsorily, then efficiency might be higher because the costs are lower; but effectiveness might be lower because the owner selling amicably might do that on condition that other conditions be changed. The changes in the plan might be achieved by using the room for discretion, or by making new or modified plans for parts of the plan area. That is likely to damage the legitimacy of the planning authority in the eyes of the public. And it is difficult to judge whether justice has been done – and, if so, in what sense.

If a situation has arisen locally whereby the citizens have become very suspicious, even cynical, about the planning policies of the local government, that government might decide to try to regain the trust of its citizens: a question of legitimacy. To do that, it might decide to have no more confidential negotiations with property owners and property developers. It might be that certain planning aims, such as the integrated redevelopment of run-down areas, cannot be achieved (a question of effectiveness) without such negotiations. In that case, the planning authority might decide:

we do not choose such planning aims until we have rebuilt our legitimacy. That can take a long time.

Adapting the Legal Approach to Social, Economic, Environmental and Physical Circumstances

Some of the choices facing a planning authority have been described above. However, a planning authority has little choice in the social, economic, environmental and physical conditions in the area which is being planned. Account must be taken of those conditions, for they affect the reactions of the citizens to planning measures – and, hence, the effectiveness of the spatial planning.

If, for example, the demand for land and buildings is high relative to the supply, suppliers will want to respond to that. A planning authority is therefore in a strong position to use the planning law in order to steer the suppliers so that they provide buildings in ways and locations that the planning authority wants. It might, for example, forbid building in an attractive area, at the same time zoning land for development elsewhere. The expectation is that the demand will be diverted to the planned area and that suppliers will want to satisfy it there. The planning authority is also in a strong position to place additional conditions on the developers, such as a contribution to the necessary infrastructure. That legal approach will be effective, also efficient (because little input will be necessary to get developers to agree), and the citizens will regard it as legitimate (output legitimacy). Its effect on economic welfare is difficult to predict: consumers might not be getting what they would prefer.

If, on the other hand, supply is greater than demand (there are, for example, vacant land and buildings) and demand is not growing, the planning authority needs the cooperation of suppliers in order to make changes, and cannot steer suppliers easily. So the planning authority might use the planning law in an adaptable way (little detail, much discretion, flexibility) so as better to meet the wishes of the suppliers. It might, if it is possible, use performance-based rather than condition-based regulations. It might even acquire land for new development; developers can then acquire building plots with much less risk than if they themselves had to buy them. In spite of that use of planning law and property rights, the effectiveness might not be high. And the necessary input might be expensive, leading to low efficiency. The legitimacy also might be low, if the planning authority is seen to be collaborating too closely with developers whose favours it wants to win.

If the context – social and economic – is changing rapidly, a different legal approach to spatial planning might be taken. For it is difficult to predict what land-use changes might be desirable, difficult also to predict the consequences of planning measures. So, planning which aims to change the land use in a particular way is likely to be ineffective, and the consequences for economic welfare unpredictable. In those circumstances, it might be better to do no more than apply certain minimum conditions which must be satisfied when changing the land use. This can be a very practical decision, irrespective of any ideological preferences. Those conditions would apply to the whole of the plan area, unless exceptions were made such as for areas of outstanding natural beauty, areas prone to flooding, areas containing valuable minerals. Effectiveness is likely to be high – the substantive spatial aims are few and it can be predicted fairly reliably how people will react to the measures. Efficiency will be high, since the only input is a bureaucratic testing of the applications against binding rules. Justice would be done, in the sense that existing property rights are not changed. And legitimacy would be high, as the actions of the planning authority would be seen as impartial and transparent.

Such an adaptation to great uncertainties can, however, be unsatisfactory under some circumstances. If, for example, the international economic or political situation is changing rapidly and unpredictably, the planning authority might not want to initiate big development projects, but neither might it want to wait passively to see how people react to a set of minimal conditions (however cleverly composed). It wants to anticipate possible changes, to be prepared. How can spatial plans be made flexible "without abolishing the desired stability and security that law and property rights create?" (Needham & Hartmann 2012: 225). One way is to create

> an institutional design that frames agenda-setting, gives incentives to market parties, allocates responsibilities and risks in a way that does not aim to find robust and long-term planning solutions, . . . but permanently to take into account the newest insights into planning challenges.
>
> (Needham & Hartmann 2012: 225)

That requires some sort of collaborative planning. Another way is to make small-scale project plans which take account of the wider geographical situation by fitting into a spatial plan which is only indicative, not binding. That uses the planning law in a particular way. Yet another way is to change the law itself – in this case, the law on property rights – so that property rights are more adaptable to changing environmental conditions, and in that way to shift the burdens of environmental change from the state (liable to pay compensation if it changes a binding plan) to the owners of those rights (Tarlock 2012).

A Planning Story

Part Two

In this second part of the planning story, what the planning authority did with the information it received (as set out in part one; see Chapter 1) is described and analysed using the framework set out above (the eight variables). But note: that framework has been set out in terms of *choices* which a planning authority can make, and different planning authorities might make different choices when faced with the same initial conditions. In order to illustrate that, the story is continued in two possible ways, each the result of different choices.

Scenario A

The local council decided formally that a land-use plan for the neighbourhood be made, taking account of the findings of the working group. The plan was to be prepared by the local officials – including planners – and any professional advisors, in close consultation with the local politicians who had special responsibility for the spatial planning of the town (the chairman of the planning committee, or similar). Those proposals were to have the form of a spatial plan with accompanying measures for realising it, and with an explanation and justification of that content. That plan would then be submitted to the local council – the elected representatives authorised to take the formal decision – for approval, possibly in an amended form.

The following land-use plan was proposed. It is described in terms of the eight (interrelated) questions raised earlier in this chapter.

Scenario A: What Are the Aims of the Spatial Planning for This Location and/or These Problems, and How Should They Be Formulated?

– The neighbourhood should remain as primarily a housing area.
– If possible, the housing should be 'affordable' for people on lower incomes. That can include students.
– The site now occupied by a cluster of derelict industrial buildings (site A) should be redeveloped for social housing and a neighbourhood park.

– The site now used for the underused industrial buildings (site B) should be made available for existing small firms, including those in the 'creative' sector.
– Shops which are or become vacant should be made available for small firms, including those in the 'creative' sector, as long as that causes no nuisance.
– Parking should be allowed only on certain streets, and then on one side of the street only.
– A system of one-way streets should be introduced so as to dissuade through-traffic.
– This plan is consistent with the existing structure plan for the whole town.

The plan specified those aims in the following detail:

– The locations now used for housing are designated as such, to prevent change of use.
– The boundaries of site A are identified precisely, but the division of that site between housing and open space is left open.
– Site B is given a designation which allows both light industry and commercial uses.
– Buildings now used by shops are given a designation which allows retail, light industry and commercial uses.
– The streets where parking is allowed are not indicated in the plan, nor the measures to dissuade through-traffic: a traffic plan for that geographical sector of the town (including the town centre) must first be made. That traffic plan should come as soon as possible, for uncertainty about the volume of traffic on particular streets can affect willingness to invest in property on those streets.

Scenario A: How to Take Account of the Existing Property Rights in the Planning Area?

– Preemption rights will be imposed on certain indicated housing areas. That would enable the planning authority to acquire housing to prevent speculation or gentrification. If the planning authority acquires housing in that way, it will offer to sell it to the local housing association (for social housing) or to the university (for student housing).
– If the existing owner of a house uses it in a way which endangers the public order (e.g. drugs dealing, prostitution), the planning authority will try to acquire the house compulsorily.

- If the existing owner of a house does not maintain it well, so that it becomes unfit for habitation, the planning authority will demand better maintenance or will close the house.
- The planning authority will acquire the land and buildings on site A. The site will be cleared and part of it offered to the local housing association (for social housing) or to the university (for student housing). The division of the site between housing and park will be determined in negotiation with the housing developer. The park will be provided and paid for by the planning authority.
- The planning authority will attempt to find one buyer for site B. If the complex cannot be exploited without a subsidy, the authority will attempt to find an external subsidiser (e.g. the European Union).
- The planning authority will act as an intermediary for existing small firms looking for new premises in the area, either on site B or in vacant shops.

Scenario A: How to Use the Available Planning Laws?

- One plan is made for the whole area.
- The planning law allows a site to be designated for more than one specified use. This is done for site B and for buildings now used for shops.
- The planning law does not allow the type of housing (cheap or expensive, to buy or to rent, commercial or social landlord) to be specified.
- The planning law does not allow 'unearned increments' in value (for example, caused by the new park) to be 'creamed off'.
- Planning law allows both compulsory purchase and preemption in certain precisely defined situations.

Scenario A: How to Take Account of Citizens' Rights?

- As soon as the planning authority received the report of the working group, it made in secret preparations for imposing preemption rights on certain locations. The announcement that such rights were to be imposed was made on the date at which those rights came into force. That legal procedure is intended to prevent speculation in anticipation of the new restrictions on selling. There is no 'transparency'.
- When a draft plan had been approved by the local council, it was published and comments invited, as is legally required.
- The draft plan described the consultations and participation which preceded the plan and how the plan had taken account of that information.
- If compulsory purchase is applied, the court ensures that the citizens' rights are respected.

- If the planning authority acquires properties amicably, it does not make public its intentions beforehand, as that could weaken its negotiating position. That is not transparent.

Scenario A: What Degree of Effectiveness and Efficiency Is Desired?

- The planning authority wants the neighbourhood to continue as a housing area. The rules will probably ensure that:
- The planning authority wants the housing to be available primarily for lower income households, including students. The effectiveness of the measures taken to achieve that will depend partly on the demand for housing: if that is high, owners of such housing can make high profits by supplying to richer people and will take all measures – legal and illegal – to avoid the pre-emption rights. And if the demand for housing is low, neither a housing association nor the university will want to acquire land or buildings to provide housing. In other words, the effectiveness of measures taken to influence the type of housing in the area might be low.
- The planning authority wants a particular development for site A. It assumes that the existing owners will want to sell. If they do not, compulsory purpose can be applied. The new development is then partly in the hands of the planning authority itself (the park) and that is assured. It is also in the hands of a housing association, and that cannot be assured.
- The planning authority wants small, light industry and firms in the creative sector to locate in existing industrial premises (site B) and in vacant shops. If the demand by such activities is small, that will not be achieved. The effectiveness can be increased by subsidising such premises, but even then, demand might be too small.
- The effectiveness of the traffic measures is likely to be high, for demand for parking and travelling is high, and supply of the necessary facilities is in the hands of the planning authority.
- Realising some of the aims is urgent – in particular, improving the social environment by reducing antisocial activities on the street. Effectiveness in the short run is desired. With the measures included in the plan, this cannot be guaranteed.
- The efficiency of the proposed measures can be expressed in terms of the (financial) costs and benefits to the planning authority. With the plan as proposed, that government has mainly procedural costs (apart from the new park). If it should be necessary (in order to realise the aims) that the planning authority acquires land and buildings, the

costs will be much higher. The returns depend on the disposal costs of the land and buildings thus acquired. Both costs and returns depend on the state of the real estate market: if demand is low, costs and returns will be low; if demand is high, costs and returns will be high.

Scenario A: How Much Importance to Give to Economic Welfare?

– The land and buildings in the neighbourhood contribute to the economic welfare of the people living there. The fact that, in this case, the plan has the approval of most of those residents indicates that they are satisfied with that contribution (although that does not mean that the residents would not enjoy higher welfare with a different plan).

– The land and buildings in the neighbourhood also contribute to the economic welfare of the whole town. This plan does no more than place the neighbourhood plan within the structure plan for the whole town. That reticence is understandable, for it is exceedingly difficult to say anything about the contribution a part of an urban area makes, or can make, to the economic welfare of the citizens in the whole urban area. The assumption is that the planning authority takes this into account, implicitly or explicitly, when making a structure plan.

Scenario A: What Interpretation of Justice to Use?

– In this case, the planning authority interprets justice in the following way: the land-use plan should respect the existing rights and benefits, but where there are conflicts, those should be resolved in the interests of the local residents, especially those who have lived there a long time.

Scenario A: How Are the Planning Actions to Be Legitimised?

– In this case, the planning authority has made clear – both by setting up a working group which includes the local residents, and in the discussions with those residents and others – that it values the trust and support of its citizens (wherever they live) and wants to retain that. The application of compulsory purchase, and even more of preemption rights, is often socially contested. But as long as most citizens regard the outcomes of such actions as being in the interests of a wider public, those actions do not undermine the legitimacy of the planning authority.

Scenario B

Shortly after the working group had submitted its report, a new local council was elected. A coalition was formed of parties which together had a majority of the seats on the new council. After a few weeks of internal discussion within the coalition, the council announced that it was not going to make a new land-use plan for the neighbourhood. The reasons given were: the problems in the neighbourhood were not pressing, and the council wanted to spend its attention (political and professional) and money on areas and projects where the problems were greater. Moreover, it was pointed out that there was 'market interest' in some parts of the neighbourhood; perhaps that alone would lead to improvements there.

The head of the planning department told the councillors in the majority coalition that the local residents had been led to expect that their contribution to the report of the working group would result in positive actions to improve their neighbourhood. They would be very disappointed. The answer: the council decides, there is a new council and it is not bound to the policies of the old council.

When the local newspaper printed a series of articles on illegal practices in the area – drug dealing, prostitution, overcrowding and dangerous living conditions in some rented dwellings – the local council announced that it would intensify police surveillance there. If necessary, houses would be vacated forcibly and boarded up as 'unfit for human habitation'.

In response to further protests by the local residents about their living conditions, the planning authority approached a housing association which was improving rental properties elsewhere in the town. The result was that residents in the neighbourhood would be given priority when the improved dwellings were re-allocated.

The neighbourhood went slowly downhill, and more and more properties – housing, shops, workplaces – became vacant. The owners were required by the planning authority to board up the properties which had been vacant for more than six months, to prevent illegal occupation. Order was maintained on the streets and other public places by good policing. More and more cars were parked on the streets, to the benefit of those using the nearby town centre. And when a commercial provider of parking asked for permission to use vacant industrial land for car parking (the owner of that land was only too pleased to rent it for any use), the planning authority gave temporary planning permission.

Then, 'sold' notices began to appear on some of the properties where previously 'for sale' notices had hung. Journalists from the local newspaper asked the various estate agents involved who the buyers were, only to be told that the information was confidential. For a long time, the sold properties remained empty. Then the council announced that it was preparing a new land-use plan for the area: the local university wanted to use it for a second campus, and the council supported that wish fully.

Later, it became known that the mystery buyer of the properties in the neighbourhood was the university: it had worked through several estate agents so as not to drive up the prices. And the university had, informally, told the leaders of the local council of its intentions. Those leaders had given their support in principle, and both parties had agreed not to spread the information further.

Work started on preparing a new land-use plan. The whole area was to be designated 'educational uses', which would allow also student accommodation and cafés. When the new plan had been approved, it would give legal justification for expropriating properties if that should be necessary.

This spatial planning practice is now described in terms of the eight (interrelated) questions raised at the beginning of this chapter.

Scenario B: What Are the Aims of the Spatial Planning for Thus Location and/or These Problems, and How Are They Formulated?

– Initially, the explicit aim was that public health and safety be maintained in the area. The implicit aim was that the existing uses should remain.
– Subsequently, the aim was that (most of) the area should be used for an extension of the local university, accommodating housing for teaching and research, for the necessary administration, for student residences, for eating and sports facilities, etc. The new land-use plan gave no more detail, except for how the campus would be connected to the road network. The university itself would make the detailed land-use plan. All building plans would be required to satisfy the building regulations.

Scenario B: How to Take Account of the Existing Property Rights in the Planning Area?

– Initially, the planning authority did nothing which would affect people's property rights, except when the exercise of those rights endangered public health and safety. It is true that in very many cases the value of property rights had declined because of the inaction of the planning authority. But no one has the right to expect a state body to make a spatial plan that will ensure that property rights do not decrease in value.

– Subsequently, many of the property rights had been acquired by the university, which wanted to use them for educational purposes, and that would be permitted. It could not be expected of the other owners of property rights in the area that they would want to use them for educational purposes, and if they did want to, that was infeasible, as the proposed layout of the campus was not in line with the existing ownership boundaries (except, perhaps, for the owner of a café in the area). So those other owners could choose between selling amicably to the university, or being expropriated by the planning authority. If they chose the latter, they would probably receive less than by an amicable sale.

Scenario B: *How to Use the Available Planning Laws?*

– Initially, the planning laws could have been used to prevent someone changing the land use to one which differed from the existing land-use plan. In practice, there was very little desire to do that. The exception was the wish to use industrial land for car parking – that was allowed by granting temporary permission. Planning laws, in the wider sense of ordinances and urban codes, were used in this first stage to prevent buildings and spaces being used in undesirable ways.

– Subsequently, a new land-use plan was made. The designation 'educational uses' could have been realised by any developer, and if one had applied for planning permission, that application would have had to be granted. In practice, the local university owned so much of the land that it was the only feasible developer. The plan was general, and gave the university a great deal of freedom to fill in the details.

Scenario B: *How to Take Account of Citizens' Rights?*

– Initially, the only citizens' right that was relevant – no new plan was being made, very few planning permissions granted – was the right to be heard. That right had been fully respected and the local residents had been able to make their wishes known. That their wishes had not been fulfilled is not a denial of that right. Nor had the 'principles of responsible government' been transgressed.

– Subsequently, all citizens with a recognised interest would have the right to be involved in the procedures for making and approving the new land-use plan. However, by that time very many of the previous residents (all of them 'interested parties') had left the area.

Scenario B: What Degree of Effectiveness and Efficiency Is Desired?

– Initially, the planning authority aimed for no effect other than public health and safety. For that to be achieved, local policing has to be intensive. That is expensive. Redeveloping the area so as to create safe public spaces would have achieved the same effect. It would have cost much more (but would have provided many additional benefits).

– Subsequently, the planning authority aimed for a new university campus in the town. It was almost certain that this would be realised. The only cost to the planning authority was making and enforcing a general land-use plan.

Scenario B: How Much Importance to Give to Economic Welfare?

– Initially, the use of land and buildings in the neighbourhood produced little economic welfare. If the area had been redeveloped, it would have contributed more economic welfare to the town.

– Subsequently, as a university campus the area would produce more economic welfare. Perhaps another use would have been a better use of the resources of land and buildings there, but it is difficult to make any general statement about that possibility.

– It is in theory possible – by using discounting techniques (see Chapter 6) – to compare the economic welfare produced by keeping the lower land use initially and then changing to the higher land use, with the economic welfare produced by a higher use introduced initially and retained.

Scenario B: What Interpretation of Justice to Use?

– The planning authority interprets justice implicitly, saying that as long as land and buildings are held and exchanged legally, then justice is done. There were no attempts, initially or subsequently, to achieve a different distribution of the benefits to be realised from the land and buildings.

Scenario B: How Are the Planning Actions to Be Legitimised?

– When the new local council decided not to make a new land-use plan, it acted legally, but it breached the trust that had grown between the local citizens, the local council and the local officials.

– When the planning authority heard of the intention of the local university to acquire land and buildings, with a view to getting the planning authority to agree to make a new land-use plan which would allow a new campus to be built, the planning authority kept that information secret so as not to weaken the bargaining position of the university when acquiring the necessary properties. The planning authority was not acting illegally in withholding that information from the public, but once again its actions would be seen as not legitimate by many members of the public.

References

Needham, B., Hartmann, T., 2012. Conclusion. In Hartmann, T., Needham, B. (eds.), *Planning by law and property rights reconsidered* (pp. 219–227). Farnham, UK: Ashgate

Tarlock, A. D., 2012. Global climate change and the stability of property rights. In Hartmann, T., Needham, B. (eds.), *Planning by law and property rights reconsidered* (pp. 135–156). Farnham, UK: Ashgate

Index

actions (of a state body): administrative
  16, 71; legislative 71, 79
active planning *see* legal approach
administrative actions (of a state body)
  see actions
administrative law *see* public law
aims (of spatial planning): 1, 12–13, 20,
  87, 89, 93, 148, 149
allocative efficiency (of use of economic
  resources) 97–8, 113

Bentham, Jeremy 128
building codes 48, 133
building regulations 14, 48, 55
bundle of rights (in land and buildings)
  see ownership

case law *see* law
citizen (and the state) 1, 12, 14, 15,
  19, 56, 63, 69 et seq., 128, 139,
  142, 149
citizen's rights: in codified or in common
  law 81; in international treaties 80; in
  jurisprudence 81
civil law *see* private law
civil rights *see* citizens' rights
club (goods) *see* goods
codified law *see* law
collaborative planning 142, 154
collective goods *see* goods
common law *see* law
common pool resources *see* resources
compulsory purchase *see* expropriation
condition-based (regulations) *see*
  regulations
conformance *see* effectiveness of spatial
  planning

contract law *see* law
cost–benefit analysis 104–5, 128, 143

democracy: direct 139; liberal 15, 19, 20,
  55, 69 et seq.; representative 139
detail (in spatial planning) *see* spatial
  planning
discretion (in spatial planning) *see*
  spatial planning
distribution of costs and benefits: from
  property rights 30–1, 33; in spatial
  planning 58, 126
distribution of welfare (effects on market
  prices) 104, 120
distributional justice *see* justice
doctrine of estates 35
due process (in law) 14, 70, 80–1, 142
duration (of permitted use) 37

easements 14, 27, 37
economic efficiency (of use of economic
  resources) 20, 86, 98–9
economic evaluation: discounting the
  future 119, 122n17; distributional
  issues 104, 120; moral issues 118,
  122n16; political issues 119
economic optimum (of use of economic
  resources) 98–9, 106, 117, 122n8,
  122n13
economic resources: scarcity of 98;
  in spatial planning 20, 21, 97–9
economic welfare 97–9; measurement of
  99 et seq.
effectiveness (of spatial planning) 85
  et seq.; conflicting goals 91–2;
  conformance 87, 93; intervening
  variables 92; performance 87;

predictions 90–1; unintended
consequences 92
efficiency (economic) *see* economic
efficiency
efficiency of spatial planning 21, 85
et seq., 150
equality: in access to land and buildings
21, 128, 129, 132; of opportunities
129–30; of outcomes 129–30;
of rights 130
ethical code 146
evaluation (of spatial planning rules) 86
ex ante evaluation *see* evaluation
ex durante evaluation *see* evaluation
ex post evaluation *see* evaluation
excludability (of use of economic
resources) 103, 107–9, 112, 122n12
executive actions (of a state body) *see*
actions
expropriation 18, 36, 48, 54, 58–9, 143
external effects (economic) 51, 93,
100–1, 103, 105, 108, 111, 114
external goals *see* goals (of policy)

fairness *see* justice
financial compensation (for planning
actions) 56, 58–60, 99, 127, 134–5
flexibility (in spatial planning) *see* spatial
planning

goals (of policy) 20, 86, 91–3
goals (of spatial planning) *see* aims of
spatial planning
goods: club goods 103, 112; private goods
103, 121n2; public goods (collective
goods) 103, 112, 121n2, 122n12;
spatial primary goods 133
governance 65, 141; good governance
70, 138, 146

horizontal co-ordination 56–7, 149

impartiality (in spatial planning) *see*
rules (of a state body)
imperfect information (and transaction
costs) 112, 115
instruments (of land and/or spatial
policy) 14, 16–17, 44–5, 48–9, 85–6
interested parties (in law) 78–9, 82
interests (in land and buildings) 15, 25
et seq., 70, 125; protection of 26–8;
value of 29–30
internal goals *see* goals (of policy)

judicial review 71, 79
jurisprudence 53, 81
justice 125 et seq.; in acquisition 134;
distributive 60, 125; ecological 129;
egalitarian 129–31; entitlement
theory 134; framework 133–5;
intergenerational 129; poverty
absolute 132; poverty relative 132; in
spatial planning 20–2, 126, 135;
sufficiency 131–3; in transfer 134;
utilitarian 128–9

land-use plan 17–9, 23, 33, 49, 51–2, 56,
71, 142
land-use planning *see* spatial planning
law: case 32, 81; citizens' rights in law
see citizens' rights; codified 32, 52–4;
common 32, 52–3; contract 15, 18,
88, 111; planning 48 et seq.; private
15, 18, 29, 44–5, 106, 107, 110–11,
134; property rights in law *see*
property rights; public 15, 18, 29–30,
43–5, 49, 69, 107, 113; statute 32,
69, 81
lease 14, 27, 37–9
leasehold *see* lease
legal approach to planning 17–23, 88,
90, 105, 126, 143: active 18; passive
18, 91; regulation 113–16; structuring
110–13
legal certainty 60, 63, 74, 93–4
legal instruments *see* instruments
legal person 14, 18, 26, 29, 38–9, 41–2,
49, 81, 110
legality 79, 140
legislative actions (of a state body)
see actions
legitimacy 138 et seq.; forms of 141;
input 141; output 142; in spatial
planning 22; throughput 142
locational monopolies *see* monopolies
locationally generic (rules) *see* rules for
land use
locationally specific (rules) *see* rules for
land use

market failures 97, 107–10; corrected by
regulating markets 113–16; corrected
by structuring markets 110–13
market imperfections *see* market failures
markets 18, 51, 97, 106–7; market
outcomes 100–3, 117; market prices
100–104; market quantities 100–3

material flexibility *see* spatial planning
measures (planning) 16–17, 45
missed opportunities *see* transaction costs
monopolies 108; locational 101, 115,
    127; natural 101–2, 115

negative external effects *see* external
    effects
NIMBY 127
nomocratic *see* public regulation
non-market values 103–4
Nozick, Robert 130, 133, 134, 136n5

order (in land use) 49–52; cosmos 50;
    designed 50; spontaneous 50; taxis 50
Ostrom, Elinor 40
ownership (of rights in land and
    buildings) 35–6; absolute 35, 43;
    abusus 36; bundle of rights 35;
    property owners 38; usus 36; usus
    fructus 36

Pareto optimum (of use of economic
    resources) 99
passive planning *see* legal approach
perfect market 106, 110
performance-based (regulations)
    see regulations
permissive planning *see* legal approach
planning authority 13, 19
planning law 48 et seq.; justification of
    measures 55–6; land-value changes
    58; prescribed procedures 58, 78;
    what is regulated 55–6
planning rights 70, 76, 130
polder model, Dutch 139
positive external effects *see* external
    effects
poverty *see* justice
preemption right 14, 45, 48, 54
private goods *see* goods
private law *see* law
procedural flexibility *see* spatial planning
property owners *see* ownership and *see*
    legal persons
property regimes 38, 41–2
property rights 13, 14, 25 et seq.;
    ambiguities in 111; customary 33;
    exercised jointly 39–40; formal 31;
    full 35–36; incomplete 111;
    indigenous 34; informal 31;
    limitations on exercise of 43–4; partial
    35–6; personal (in persona) 39–40;

presumptive 34; real (in rem) 39–40;
    as a social relationship 30–1; transfer
    of 39–40
proportionality (in rules of a state body)
    see rules of a state body
public goods *see* goods
public interest 28, 54, 76–7, 83n3,
    122n11, 138–9, 142
public law *see* law
public regulation: nomocratic 51–2, 134;
    teleocratic 51–2

rational justification *see* rules of a state
    body
Rawls, John 16, 133, 136n5
regime (of land laws) 45
regime (of user rights) *see* regime
    (of land laws)
regulation (planning by) *see* legal
    approach to planning
regulations: condition-based 63–5, 153;
    performance-based 63–5, 153
regulatory failures 116
regulatory planning *see* legal approach to
    planning
resources 97–8, 121n3 and *see* goods:
    common pool 40; open access 103,
    112; scarcity of 98
right to be heard 76
rights in land and buildings *see* property
    rights
rivalry (in use of economic resources)
    102, 109
rule of law 55, 63, 72
rule systems 49, 51–2, 88
rules (of a state body) 73; impartiality
    75; proportionality 75–6; rational
    justification 75; stability 34, 73–4, 154
rules 15–16; administrative 16; civic
    society 16; constitutional 16;
    legislative 16; pre-constitutional 16
rules for land use 52, 88; design 51;
    framework 51; locationally generic 45;
    locationally specific 45, 127; map-
    dependent 52; map-independent 52

scarcity *see* resources
Sen, Amartya 132
servitudes 32
spatial order *see* order
spatial planning: activity of 12–13;
    degree of detail 61; discretion 62, 90;
    flexibility 61, 62–3, 90

stability (in rules of a state body) *see*
  rules (of a state body)
statute law *see* law
structuring (planning by) *see* legal
  approach to planning

taking (legal, of property rights) 59
teleocratic *see* public regulation
termination (of permitted use) 37
third party interests *see* interested
  parties
transaction costs 86, 109, 111, 115–16;
  missed opportunities 50, 112

trias politica 15, 79
trust (of the citizen in the state) 139

ultra vires 72, 76
urban code 19, 45, 49–52
utilitarianism 128–9

value capturing 60, 127, 131
vertical co-ordination 57, 149

welfare (economic) *see* economic welfare

zoning plan 52, 56, 93